Unraveling Evolution

Second Edition

With 170 new photos and 12 new appendices

By Joshua Gurtler

Truth
Publications

Taking His hand,
Helping each other home. ™

ISBN 10: 1-58427-416-6

ISBN 13: 978-1-58427-416-2

First Edition: 2006

Cover photo: istockphoto.com

Truth Publications, Inc.
CEI Bookstore
220 S. Marion St., Athens, AL 35611
855-492-6657
sales@truthpublications.com
www.truthbooks.com

Acknowledgments

From the 1st edition, I would like to express my deep appreciation to Karen Witherington for proofreading the original manuscript for grammar, style, and consistency. Mike Willis put many long hours into formatting and revising the photographs and text of the manuscript. Steven J. Wallace spent a great deal of time reviewing the manuscript and added many helpful comments and additions. Thank all of you. The following individuals also deserve special recognition for reading the original manuscript for me. Rob Bradshaw (biblicalstudies.org.uk), David Dann, Dr. Larry Dickens (Former Professor of Chemistry, Florida College), John E. Gurtler, Dr. Brad Harrub (Ph.D. Neurobiology and Anatomy) Dr. Chyer Kim, P. W. Martin, Leon Mauldin, Joshua Reaves, Dr. Steven N. Renfrow (Physicist, Scientic, Inc. Aerospace/Defense), Dr. Glenner Richards (Professor of Biology, Evangel University), and Dr. Peter Taormina Director of Science, John Morrell, Inc.

I would also like to thank a number of individuals who were especially helpful in granting me permission to reproduce their copyrighted illustrations and/or photographs or in assisting me in acquiring permission. These individuals are Dr. Michael J. Behe (Professor of Biochemistry, LeHigh University), Vera and Dina Cheal, Roy Cripps, Dr. Donald Forsdyke (Professor Emeritus in Biochemistry, Queen's University), Dr. Allen Glazner (Professor of Geological Sciences, University of North Carolina), Charles McCown (along with Dr. Bert Thompson from Apologetics Press), Johan Opsomer, Don Patton, Chet Robbins, Steve Rudd, Jody F. Sjogren, and Bob West. (Please note that any assistance provided by the individuals mentioned above does not necessarily indicate our total agreement in all biblical, theological, or apologetics-related issues.)

Dedication

I would like to dedicate this work to the late R.C. Hammonds and his wife, Sibyl Hammonds, my beloved grandparents, as well as to my late uncle, Richard Copeland. Your humble and devout influences extend farther than you may ever know.

Biography

Joshua Gurtler was born to John and Joan (Hammonds) Gurtler in 1974. He is married to Jana (Godwin), also from Alabama, and they have three children, Tristan, McKenna, and Landon. Joshua has preached the gospel in the U.S., three foreign countries, and for the Eastside church of Christ in Newnan, GA from 1999-2007. Truth Publications has published his workbook, *Unraveling Evolution,* suitable for teenagers or adults, since 2006. Joshua studied at Florida College, Wallace State Community College, Auburn University, and the University of Georgia, where he earned his Ph.D. in food microbiology in 2006. He conducted post-doctoral research in 2006 and 2007 at UGA, while also being employed by a food safety consulting firm. Joshua has lectured on food microbiology issues in the U.S., Canada, South Korea, and China and has published numerous book chapters and peer-reviewed articles in scientific journals, in addition to editing three books pertaining to food microbiology. He has served on graduate advisory committees for M.S. and Ph.D. students, has secured numerous grants or funded research agreements and is currently a scientific co-editor or associate editor for two scientific journals. From 2007 to 2018, Joshua has been employed as a research scientist for the USDA, Agricultural Research Service, Eastern Regional Research Center in Wyndmoor, PA.

Can We Trust the Bible As the Inspired Word of God?

Introduction

There is an indispensable belief one must have prior to cultivating an understanding of true ethics (a moral value system). This belief also precedes understanding the truth regarding creation versus humanistic evolution—a belief in God himself. Debating biblical morality with an atheist is akin to arguing integral calculus with an individual who denies that 1+1=2. The aim of this first lesson is to provide undeniable evidence, to the rational mind, that there is a God who lives and moves and created the universe we inhabit. Before we delineate these arguments, let's first see what this world would be like if everyone were to believe there was no God.

The *Humanist Manifesto II* said that *"Ethics is autonomous and situational."* That is to say, ethics is decided by each person in each given situation. There are no moral absolutes.

I. What if There Were No God? (The Consequences of Atheism)

To assume there is no creator necessarily means that there is no divine authority, which in turn means no higher standard exists to which we can appeal for ethical decisions of right and wrong. Each man would be a law to himself. *The Humanist Manifesto I and II,* serving as the atheist's Bible states, *"Ethics is autonomous and situational"* (1977, p. 17). That is, morality is dependent upon any one person's perception of right or wrong in any given situation. There would be no universal standards for mankind and neither could any

Crematoria in Hitler's extermination camp in Dachau, Germany (wikicommons).

one individual's decision be considered good or bad. The atheist cannot say it is wrong to commit murder, genocide, rape, abuse animals or the environment, torture or maim other human beings or enslave people against their will. Atheists can say they don't "like" or "approve" of such actions, but if there is no God, then condemning these acts is a philosophical impossibility.

The German philosopher and avowed atheist Friedrich Wilhelm Nietzsche (1844-1900) said, *"morality is a hindrance to the development of new and better customs: it makes stupid"* (1881, s. 19). Adolph Hitler, a committed student of Nietzsche, followed this philosophy to its logical end, abolished morals, and murdered up to 10 million *"undesirables."* The only rational conclusion, if there is no God, would be that no action is any better or worse than any other and we, as the Apostle Paul said, *"are of all men the most pitiable"* (1 Cor. 15:19).

Harvard professor of entomology, two time Pulitzer Prize winner and avowed atheist, Dr. E.O. Wilson (Harrison). Dr. Wilson told me he became an atheist when he went to the University of Alabama and was taught evolution.

Paul Kurtz. One of two authors of *The Humanist Manifesto II* (wikicommons).

Friedrich Wilhelm Nietzsche's atheistic philosophies emboldened Hitler's genocidal pogroms.

Dr. Richard Dawkins, the world's greatest spokesman for atheism, said *"What's to prevent us from saying Hitler wasn't right? I mean, that is a genuinely difficult question"* (Dawkins, 2007) (Photo: Dawkins, 2009).

Adolph Hitler drew heavily from the atheistic teaching of Nietzsche and the evolutionary teaching of Darwin to justify his pogroms of mass genocide.

Oxford Professor, Dr. Richard Dawkins, is the foremost spokesman for atheism today, and very hostile against the Bible and Jehovah. Dawkins was a believer and a member of the Anglican church until he was taught evolution in school and became an atheist. In 2006 he published, The God Delusion, for which there have been numerous responses.

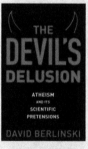

Dr. David Berlinski (in 2009) responded to Dawkins' The God Delusion, with his book, The Devil's Delusion: Atheism and its Scientific Pretensions. Professor Berlinski has taught at 10 universities, and is also a Discovery Institute Senior Fellow.

Numerous quotes from leading atheists could be provided, but the following two quotes from three of the most renowned agnostics should suffice: "We humans are modified monkeys, not the favored Creation of a Benevolent God… As evolutionists we see that no (ethical) justification of the traditional kind is possible… In an important sense, ethics as we understand it is an illusion" (Ruse and Wilson, 1993). If there is truly no "(ethical) justification" then how can we say murder or even the holocaust was wrong? This very question was posed to Richard Dawkins. In an interview, Dawkins was asked, "If we do not acknowledge some sort of external [standard], what is to prevent us from saying that the Muslim [extremists] aren't right?" And Dawkins response? "Yes, absolutely fascinating. What's to prevent us from saying Hitler wasn't right? I mean, that is a genuinely difficult question" (Dawkins, 2007).

II. God and Science

For the balance of this lesson we will present evidence attesting to the existence of God. One might argue, however, "you can't prove God by science." We counter that it is impossible for anyone to disprove God by science. Steven J. Wallace stated, "to prove that something doesn't exist, you would have to observe everything everywhere at the same time—in short, you would have to be God!" (2005). Let us consider the argument that God's existence cannot be proven scientifically. Although empirical (experimental) or scientific evidence is highly regarded, it is not without flaw. The scientific method requires, at a minimum, (1) Observation of some event in the universe, (2) Development of a hypothesis to explain this event, (3) Utilization of the hypothesis to design experiments that will prove or disprove the hypothesis, and (4) Conducting repeatable experiments to validate the hypothesis. Thus, anything that cannot be repeated in experimental trials cannot be "scientifically proven." We cannot "prove," by scientifically reproducible experimentation, that Alexander the Great, or the Pharaohs, or the Caesars actually lived, but we believe that they lived based on other undeniable and just as credible evidence (e.g., archaeological and historical findings). In the same way, there are many ways to demonstrate that God exists without putting him in the scientific test tube, so to speak.

But if there is no God, where did the first life come from? Evolutionist Dr. Derek Ager said, "Evolutionary accidents have furnished and finished many, many species in the past and certainly more will go in the future… The most unlikely accident of all in the history of life on Earth was the origin of life itself. Someone has compared the likelihood of the right chemicals and the right forces coming together in the primaeval soup to the likelihood of a hurricane passing through a junkyard and blowing the pieces together to form a jumbo jet" (1993b, pp. 139, 149).

III. Evidence for the Existence of God

Numerous arguments postulated over the years have demonstrated that there must be a higher force or a greater being at work in our world and in our lives. We will examine only five arguments for God's existence; however, we encourage our readers to investigate many other arguments for God's existence.

Renowned atheistic debater and former Oxford Professor, Dr. Antony Flew converted to theism in 2004, strongly influenced by arguments from "Intelligent Design."

A. First Cause (the Cosmological Argument). It is an axiom (a self evident truth) that every event or effect has a cause; and the cause of the effect has a cause; and the cause of the cause of the effect has a cause, and so on. But, have you ever tried to follow that line of reasoning back to its logical end... or beginning? It's rather mind boggling isn't it? Eventually you must conclude that there was a **first** cause from which all subsequent causes originate. The Hebrew writer states this much in Hebrews 3:4, "For every house is built by someone, but He who built all things is God." That is, everything was created by something, **except for God!** God's name itself, Jehovah or YHWH in Hebrew, means the existing one. He has always been, without beginning, and will always be, without end. This is a concept too difficult for the human mind to fully comprehend. Yet, how could there have ever been nothing in existence since there is something now? Notice the following quote, "In this connection, let us think about this: if there was ever a time when absolutely nothing existed, then there would be nothing now, for nothing produces nothing but nothingness! Since something does exist, it must logically follow that something has always existed. Exactly what was that?" (Thompson and Jackson, 1992a, p. 2).

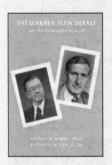

Dr. Antony Flew (while still an atheist) once debated church of Christ evangelist and College President, Dr. Thomas B. Warren, on the existence of God at North Texas State University in 1976 (see the book, The Warren-Flew Debate, at CEIbooks.com or other bookstore. The audio of the Warren-Flew debate is available on YouTube).

There are, then, only three options for the origin of the universe:

1. The universe has always been. This is not a viable option because of the **second law of thermodynamics** (the law of increasing entropy). This law states that everything is moving from a more organized and orderly state toward a state of less usable energy, confusion, randomness, chaos, and disorder. That is, if the universe was eternal, it would be a chaotic mess and not the orderly nature of orbiting planets, moons, comets, solar systems and stars that we see. Even modern science has conceded that the universe must have had a beginning. The famous Dr. Edwin Hubble, in 1929, plotted the speed of galaxies and determined that they are moving apart from one another and receding from the earth, resulting in the modern theory of the "expanding universe." The scientific community has almost universally accepted this argument, admitting that our universe had a beginning. What caused it, they don't know.

2. The universe created itself from nothing. This neither is a reasonable conclusion based on the **first law of thermodynamics** (the law of the conservation of matter) that states that matter can be neither created or destroyed. It is physically, scientifically, philosophically, and rationally impossible that something could come from nothing. Still curious? We challenge you to ask any scientist, physicist, philosopher, or logician if matter can spontaneously arise from nothing.

3. The universe originated from a first cause. The only logically-acceptable conclusion

Dr. Edwin Hubble discovered that the universe had a beginning.

Renowned NASA physicist Dr. Robert Jastrow, who, although agnostic, admitted to finding God in science (nss.org).

"The Case for the Existence of God" presents some of the strongest historic arguments for God. This 207-page book is available for purchase or FREE in PDF format from www.apologeticspress.org.

is that the universe was created by a supernatural (metaphysical) force. In fact, after spending decades attempting to discredit a creator, scientists have been reaching this conclusion in mass: "For the scientist who has lived by his faith in the power of reason, the story ends like a bad dream. He has scaled the mountains of ignorance; he is about to conquer the highest peak; as he pulls himself over the final rock, he is greeted by a band of theologians who have been sitting there for centuries" (Jastrow, 1992 , pp. 106, 107). Thus God, the uncreated and eternal, created all. "Hence to ask who made God would be like asking who made the unmakable being. To ask who made a necessary entity is to talk nonsense" (Hoover, 1992, ch. 6). "Is it not unreasonable to believe that from 'nothing' came 'something' and from 'lifeless matter' came 'living matter'? When would 'nothing' produce 'something'? Can you imagine how out of 'nothing' came 'lifeless matter' which then somehow produced 'reason'?" (Wallace, 2005).

B. Design (the Teleological Argument). Failing to see handcrafted design in our surrounding environ is indeed one of the saddest and most myopic mistakes one could ever make. The Scripture attests to God's design in the natural realm through what many call "natural revelation" (Ps. 19:1-6; Acts 14:15-17; 17:22-31; Rom. 1:18-20).

1. Design of the universe. The celestial heavens and bodies are a testament to great thought, work, and design. The Earth's orbit departs from a straight line by only 1/9 of an inch every eighteen miles. If it departed 1/8 of an inch, the sun would incinerate us; and if it departed by 1/10 of an inch, we would freeze (Gribbin, 1983, pp. 36, 37, 40, 102). The earth is 240,000 miles from the moon but if it was only 1/5 of a mile closer, the moon's gravitational pull would cause the ocean's tides to cover the continents with 35-50 feet of water twice a day (Thompson and Jackson, 1996, p. 12). The moon is receding from the earth at a rate of about 3.8 cm/yr. (1.5 inches/

yr), showing that the moon and earth could not be 4.54 billion years old (creationwiki.com). If our planet's atmosphere was any thinner, the Earth would be destroyed by meteors. Ninety percent of the world's oxygen comes from oceanic aquatic plants; thus, if our oceans were smaller, or if the earth's thin crust (which contains minerals that are oxidized by and bind up oxygen) were even 10% thicker, we would soon run out of free usable oxygen (*Ibid.*).

2. Design of the human body. The renowned evolutionist Dr. Carl Sagan estimated that the information contained in the DNA in only one of our body's 100 trillion cells is around 10^{12} bits. In simpler terminology, one microscopic human cell has more information stored within it than the world's largest library of over ten million volumes (*Ibid.*, p. 14). Consider the staggering complexity of the human body. *"Hair has several functions. It is part of the body's sentry system. Eyelashes warn the eyes to close when foreign objects strike them. Body hairs also serve as levers, connected to muscles, to help squeeze the oil glands. Hair acts as a filter in the ears and nose. Hair grows to a certain length, falls out, and then… is replaced by new hair. Hair is 'programmed' to grow only to a certain length. But who provided the 'program'?"* (*Ibid.*, p. 17).

The eye is more advanced than any camera man has ever created. Charles Darwin admitted that it seems absurd to believe that the eye could evolve on its own. "To suppose the eye with all its inimitable contrivances for adjusting the focus to different distances, for admitting different amounts of light, and for the correction of spherical and chromatic aberration, could have been formed by natural selection, seems, I freely confess, absurd in the highest s*ense"* (1872, p. 143). Light images enter the eye at approximately 186,000 miles per second through an adjusting diaphragm (the iris) and is inverted upside down by the lens onto the retina at the back of the eyeball where the image is picked up by 137 million nerve endings that send the signals to the brain at over 300 miles per hour for processing (Thompson and Jackson, 1996, p. 18). The eye can handle 1.5 million messages simultaneously and gathers 80% of the information that our brain processes, and moves approximately 100,000 times a day. Our nervous system is highly advanced with more than 3-4 million skin pain sensors, 500,000 touch detectors and over 200,000 temperature gauges (*Ibid.*, p. 22). The noted evolutionist, John Pfeiffer even stated that the nervous system is, *"the most elaborate communications system ever devised"* (1961, p. 4). If so, then who devised it? Random chance? Or, an intelligent designer?

C. Animal Instinct. Instinct is the innate guiding inclination in animals by which they are programmed to eat, hunt, migrate, flee danger, and reproduce. Without instinct, animals would be extinct. Consider this. If animals had to **develop** instinct, they would have died off long before they were able to carry out functions

Gray Whale, which migrates 6,200 miles a year (NOAA, wikicommons).

Herring Gull chick in nest with egg (Haslam).

A hen, by instinct, incubating a clutch of eggs. (Photo by Johan Opsomer. Used by permission.)

necessary for life – functions that can only be accomplished through instinct! The chicken lays one egg approximately every 0.8 days. After copulation, the hen is able to store sperm in an inner cavity to fertilize each egg laid over the next few days. After laying a clutch (5 or 6 eggs) the hen somehow *knows* to incubate the eggs by sitting on them for exactly 21 days at a temperature of approximately 105ºF at which time they hatch. How does the hen know to do this? What if the alleged evolving hen decided to sit on the eggs for 15 days instead of 21? There would be no more chickens.

Gray whales migrate approximately 6,200 miles per year. They **know** to take the same path from the Arctic Ocean through the Bering Straits to the Pacific Coast of America and the Baja Peninsula

Dr. Greg Bahnsen, believer, apologist and debater, who mastered presuppositional apologetics and the transcendental argument for God in debates with atheists (youtube.com).

of California and reach San Diego California at Christmas every year to birth their calves (Gitt and Vanheiden, 1994, pp. 39, 40). How do they know to do this? The 1977 Nobel Prize winner, Albert Szent-Gyorgyi, studied feeding habits of herring gulls and their chicks. When hungry, young herring gulls somehow know to peck at a small red spot located on its parent's beak. This action, "involves a whole series of most complicated chain reactions with a horribly complex underlying nervous mechanism" (1978, p. 1). The end result is an unintentional physiological response by the parent gull to vomit up its food in order to feed the hungry chick. How does the infant gull know to do this? Evolutionist Gordon R. Taylor said, "When we ask ourselves how any instinctive pattern of behaviour arose in the first place and became hereditarily fixed, we are given no answer" (Taylor, 1982).

D. Universal Belief Argument. Peoples from every culture and society ever discovered in the history of the world have believed in the existence of a divine creator. In fact, the earliest religious beliefs discovered, apart from the Bible, may be those of monotheism and not polytheism (Jackson and Thompson, 1996, p. 16; Nelson and Broadberry, 1994).

E. Anthropologic Argument (the Moral Belief Argument). Mankind is the only creature that has ever held an innate sense of right and wrong, otherwise known as morality. It is universally accepted as wrong to kill, steal, rape, swindle, or abuse one's fellow man. Even atheists commonly use such descriptors as right versus wrong, good versus bad, we should or shouldn't, we ought or ought not. From where did these innate instincts come? If there truly is no God… then neither is there right or wrong, good or bad, moral or immoral.

Carving of 17th Century scientist Robert Boyle who confirmed the universal "Boyle's law," which states that pressure of a gas decreases as the volume of a gas increases (chemheritage.org).

In such a case, the atrocities of Adolf Hitler were justifiable and in line with Nietzsche's doctrine. Paradoxically, atheism itself has a rigid set of what we might call "religious" convictions. These include faiths they firmly believe, teach, practice, and attempt to bind on others through means such as their implementation in public schools. However, if there is no God, why does any of this matter? On the other hand, if there is a God, it is only natural that there would be instilled in us all a system of instinctive ethics motivating our opposition to what we believe to be wrong, even in the atheist.

F. The Anthropic Principle. This argument states that the universe is minutely fine tuned for our existence by numerous cosmological parameters, that, if deviating even slightly, would preclude our existence. In short, the universe was made precisely for our existence. Therefore, in a random chance, happenstance, accidental, unguided material universe, cosmological laws would not exist. There is no reason that physical laws should exist and be consistently applied in all parts of an accidental universe. Nevertheless, our universe is extremely finely tuned by dozens of known unchangeable laws such as the *strong and weak laws of gravity, Boyle's law, the four laws of thermodynamics, laws of electromagnetism, laws of photonics, laws of quantum mechanics, laws of electromagnetic radiation, laws of gravitation and relativity*, etc.

Laws must come from a lawgiver, a mind, which must have preceded the material universe. God.

G. The Transcendental Argument for God (TAG). The TAG is based on presuppositional apologetics, which says that without the God of the Bible, nothing in the world would make sense. No other worldview can account for 1. Laws of logic (and the law of noncontradiction (LNC)), 2. Uniformity of Nature (Laws of nature, The preconditions of science) and 3. Laws of morality. God cannot deny himself or lie: Titus 1:2 "in hope of eternal life which God, who cannot lie, promised before time began."

Further, God Made Us in His Image and Expects Us to Think Like Him and Imitate Him (Gen. 1:26; I Pet. 1:16; Eph. 5:1). Atheism insists on materialism (the material universe is all that exists) and cannot account for the 3 laws mentioned above. Hinduism teaches the doctrine of Maya, which says everything we see is illusion, thus, we cannot know the truth. While Buddhism and Taoism deny the LNC and say contradictions can occur. The most

No one upon seeing the likenesses of four U.S. presidents carved into the side of Mt. Rushmore could attribute their creation to evolution. Intelligent design demands an intelligent creator. (Photo rights purchased from istock.com.)

German creation scientist, Dr. Werner Gitt, in his book, "In the Beginning was Information" makes a compelling case for "Information Science." He argues that "Information Science" teaches us "There is no known law of nature, no known process, and no known sequence of events, which can cause information to originate by itself in matter (Theorom 28)" (p. 106).

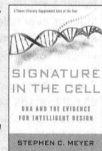

Dr. Stephen Meyer's book, Signature in the Cell makes a compelling case for intelligent design, based on the mind-boggling amount of information contained in the DNA of the cell, that must have been programmed by someone.

In The Ultimate Proof of Creation (Available in DVD, CD or book), Dr. Jason Lisle (astrophysicist) uses presuppositional apologetics to make the Transcendental Argument for God (TAG). Having been taught by Dr. Greg Bahnsen, Dr. Lisle breaks the TAG teaching down in a simple manner, using pictures, diagrams, and anecdotes to illustrate this most powerful proof for God's existence.

famous accomplished TAG apologist was the late Dr. Greg Bahnsen. I highly recommend and encourage you to listen to the Bahnsen-Stein debate on the existence of god at: www.youtube.com/watch?v=anGAazNCfdY For a fuller treatment of this subject, see **Appendix A1.** entitled, *"The Law of Noncontradiction Argues for the God of the Bible and Against Atheism, Hinduism, and Buddhism."*

Conclusion

A belief in the Creator in whom "we live and move and have our being" is the first step to understanding the truth of life's most persistent questions (Acts 17:28). For, "The fear of the Lord is the beginning of knowledge, But fools despise wisdom and instruction" (Prov. 1:7). The following articulately and poetically sums up the atheist's creed: "In the beginning nothing was. And, nothing caused man to evolve from nothingness. With the passage of time from nothingness man developed morality which suggests that we 'ought' to do right and not to do evil. But the time is coming again when

there will be nothing. Let us, therefore, fall down before the throne of Nothing and be good" (Thompson and Jackson, 1992a, p. 30).

Evolutionist, yet believer, Dr. Henry Gee (Chief Editor for the most prestigious science journal, Nature) says that religion and science aren't incompatible. (Photo credit: Huffington Post)

Dr. Henry Gee, (Senior editor for the prestigious journal *Nature*), evolutionist, yet believer said, "I am one of those people for whom Dawkins would no doubt reserve his most trenchant criticism. Dawkins thinks that science itself provides sufficient awe and wonder to replace an instinct for the supernatural. I don't. Religion, for all its ills and inequities, is one of the few things that makes us human: I am with the scientists of an earlier age, who found that their motivation in advancing the cause of knowledge was to magnify the name of the Creator" (Gee, 2006).

NOTES

Questions

TRUE OR FALSE

_____ 1. It is not necessary for one to believe in God in order to understand the truth about evolution.

_____ 2. There is no evidence for the existence of God.

_____ 3. The law of increasing entropy (i.e. the second law of thermodynamics) states that all matter is moving toward a state of disorder.

_____ 4. Animal instinct, necessary for survival and existence, could have developed through evolution over millions of years.

_____ 5. Some ancient societies that are based on atheism have been discovered.

Multiple Choice

CIRCLE ONE

1. The German philosopher who avowed that morality "makes stupid" was (A) Darwin; (B) Hitler; (C) Szent-Gyorgyi; (D) Nietzsche.

2. The Hebrew name Jehovah (or YHWH) means (A) The Mighty One; (B) The Existing One; (C) The Loving One; (D) The Horrid One.

3. The teleological argument involves (A) A first cause; (B) Philosophy; (C) Design; (D) Quantum physics.

4. Mankind is the only animal that (A) Protects its family; (B) Feels pain; (C) Communicates with its own kind; (D) Understands the concept of right versus wrong.

Short Answer

1. What does the *Humanist Manifesto II* mean when it says that *"Ethics is autonomous and situational"*? _____

_____

Are ethics autonomous and situational? _____

2. Can God be proven by the scientific method? _____ Why or why not? _____

_____

3. What is another name for the first law of thermodynamics? _____
What does this law prove? _____

_____

4. Explain the cosmological argument. _____

_____

5. Why might an atheist argue that the actions of Adolf Hitler or Josef Stalin were either justifiable or indefensible? _____

_____

In the atheist's affirmation statement that these two men may have been "right" or "wrong," what has the atheist unintentionally proven? _____

6. Under what conditions, as the apostle Paul stated, are we *of all men the most pitiable*? ____

_____

_____

Discussion Question in Preparation for Answering Unbelievers and Critics

After watching you give thanks for your meal, an acquaintance laughs and says, *"Who are you talking to? Don't you know that science has proven there is no God?"* How do you respond? ____

_____

_____

_____

_____

_____

_____

_____

_____

_____

_____

_____

Lesson 2

Is the Bible the Word of God?

Introduction

Lesson one established evidence for the existence of a Divine Creator. Although an individual may believe in God, he/she may fail to understand who God is, or the truth concerning the creation of the world. Only after one understands God's "special revelation," illuminated in His written word, will he be able to understand the truth concerning our origin.

I. Determining Truth

In almost every aspect of life, people demand truth. That is, all people desire honest, unbiased, objective answers to life's most important questions. For example, if you went to your physician with a stomach ache, would you accept the following answer? "I want you to take this medicine and see me in two months because I am convinced that your pain is either indigestion, or stomach cramps, or cancer of the pancreas." No. You would say, "I want the facts." What if a state trooper stopped you and said, "I am going to write you a citation because either you have a tail light out, or you failed to yield to a traffic signal, or you were involved in a hit and run accident." You would say, "I want the truth." When we consider an issue important, we want nothing less than the truth. However, when we turn to the most important

RECOMMENDED DISCUSSION: In the White-Crossan debate (10/29/14), Drs. James White and John D. Crossan very amiably discuss the question of "Whether the Bible is True?" Dr. White does an excellent job defending the authenticity of the Bible, despite Dr. Crossan's antibiblical arguments. Dr. White uses the presuppositional apologetic approach, discussed in Chapter 1. You can watch this debate on YouTube.

subject of all time (the origin of man, the God of man, man's eternal destiny, the nature of man's soul, and what God requires of man) instead of objectivity, people want *subjectivity*. That is, they argue that there are no absolutes. No truth. No facts. In this regard, some will argue, "Whatever you want to believe is fine. I don't want anyone changing my mind and I won't try to change his." This is called *moral relativism*. "You believe what you want to believe and I'll believe what I want to believe."

There are five major religious groups in the world. In listing these groups, we use names in the secular vernacular and not the biblical sense: Christianity, Judaism, Islam, Hinduism, and Buddhism. A careful study of the teaching of each group will find that their doctrines are diametrically opposed to one another. So, which one are we to believe? They can't all be right, can they? Judaism says there is one God, Jehovah. Christianity says that one God took the form of a man, Jesus. Islam says both are in error and should be converted or warred against if they fail to pay the Muslim tribute (*Koran,* Surah IX). Hinduism says there are thousands of gods. Buddhism says that there is no living personal God, only a powerfully-divine energy source from which we all originated and will return to some day. Either all of these groups are wrong or one is right and four are wrong, I accept the latter alternative and would like for you to consider evidence showing the Bible to be from God; evidence that the non-biblical Judaistic traditions, the Koran of Islam, the Vedic Texts and Bhagavad Gita of Hinduism, and the teachings of Buddha all lack.

II. The Unity of the Bible

Such a unified manuscript could not occur without supernatural aid. The Bible is composed of 66 books, written by 40 men from all walks of life over 1,400 years (~1,500 BC to ~ AD 100). The authors? Nehemiah was a royal cupbearer, Peter was a fisherman, Luke was

An artistic depiction of one Biblical writer, Nehemiah (ca. 446 B.C.) who was a royal cupbearer to king Araxerxes of Persia before returning to Jerusalem, to rebuild the city wall, despite opposition from Sanballat of Samaria, Tobiah the Ammonite et al. (Photo: Book of Nehemiah, 2004).

a physician, Matthew a tax collector, Solomon a King, Amos a farmer, David a shepherd, Daniel a slave and a royal servant, Paul a tent maker, Hosea the husband of a prostitute, Moses a royal prince, and Joshua a soldier. The Bible was written by rich and poor alike, in three languages (Hebrew, Aramaic, and Greek), and on two continents (Europe and Asia). Yet, unbelievably and inexplicably, only through divine intervention, the books of the Bible prove to be in perfect harmony and unity concerning the most controversial issues of all time. Among these writers, there are no disagreements, no contradictions, no correcting of each other, and, except for a very few individuals, the authors were never acquainted with one another. Biblical contradictions cannot be affirmed until all reasonable explanations are considered.

On the other hand, let's say you were to assemble 40 historians and ask them to collectively compose a single fifty-page report on the life and mind of Lee Harvey Oswald. Would there be unity? Absolutely not. Disagreements, arguments, contradictions, and errors would abound. So, how could forty men over 1,400 years write about the most controversial subject of all time without conversing or collaborating and produce the most complete, renowned, and unified historical document ever penned? The answer to such a feat is impossible except for the miraculous guiding hand of the God of the Bible. "The Bible exhibits such astounding harmony, such consistent flow, and such unparalleled unity that it defies any purely naturalistic explanation… Each book of the Bible complements the others in a single unified theme. From Genesis to Revelation there is a marvelous unfolding of the general theme…" (Thompson, 2001a, pp. 12-13).

III. The Historical Infallibility of the Bible

Other religious books or documents are typically riddled with historical and factual errors. For example, the Book of Mormon teaches that native American Indians were actually Israelites who immigrated to the Americas by boat about 2,600 years ago. However, modern genetic testing of Native Americans and Jews, by former Mormon Dr. Simon Southerton, has proven that American Indians descended from Asian blood rather than Jewish (2004). The book of Mormon also teaches that these early American Indians used wheat as a staple food. History clearly bears out that prior to the arrival of the Europeans 500 years ago, wheat had not been introduced to this continent. The following are just ten out of a multitude of ways the Bible demonstrates its historical accuracy in stark contrast to the religious books of man.

An 1843 photograph of Joseph Smith whose Book of Mormon has been proven to be a work of fiction.

A. The Bible teaches that Moses wrote the Pentateuch. The first five books of the Bible (Genesis-Deuteronomy) are commonly called the Pentateuch, which the Bible, on numerous occasions, affirms was written by Moses through inspiration (2 Chron. 34:14; Ezra 6:18; Neh. 13:1; Exod. 17:14; John 5:46; Mark 12:26). For centuries, critics have laughed and scoffed at this idea, boasting that Moses couldn't have written these books because ancient writing wasn't developed until long after Moses had died (~1451 BC). The critics were proven wrong by (A) The 1933 discovery of the city of Lachish which shows Hebrew writing contemporary to Moses (Wiseman,

Front gate of the uncovered ancient city of Lachish, mentioned in the Bible; a city which critics denied existed (Wilson).

The uncovering of Sargon's palace in 1934 in Khorsabad, Iraq silenced the critics who denied the existence of this king who was mentioned only in the Bible. Photo by Mike Willis.

1974, p. 705), (B) Phoenician writing that existed in the year 1600 BC, 150 years before the death of Moses (Thompson, *op. cit.*), (C) A table discovered at Ras Shamra containing thirty letters of Ugaritic writing in the same style as modern Hebrew, revealing that Hebrew-based writings existed at least 3,500 years ago (Jackson, 1982, p. 32), and (D) The Code of Hammurabi (dating from 2000-1700 BC) was up to 300 years older than Moses (Free and Vos, 1992, pp. 55, 103).

B. The Bible mentions Sargon II, king of Assyria (Isa. 20:1). For years, scoffers have accused the Bible of being mistaken regarding Sargon, King of Assyria who reigned 722-705 BC However, in 1843 King Sargon's Assyrian palace was discovered. Further, artifacts of King Sargon have been on display at the University of Chicago's Oriental Institute.

C. Sir William M. Ramsay attempted to historically discredit the Book of *Acts*. However, after numerous archaeological digs, Ramsay concluded that *Acts* is historically accurate. "In Acts, Luke mentions thirty-two

countries, fifty-four cities, and nine Mediterranean islands. He also mentions ninety-five persons, sixty-two of which are not named elsewhere in the NT. And his references, where checkable, are always correct. Only inspiration can account for Luke's precision" (Jackson, 1991, p. 1).

D. The historical accuracy of the New Testament. Jerry Moffit noted that "Over thirty names, including emperors, high priests, Roman governors, princes, etc., are mentioned in the New Testament, and all but a handful have been verified. In every way the Bible accounts have been found accurate (though vigorously challenged). In no single case does the Bible let us down in geographical accuracy. Without one mistake, the Bible lists around forty-five countries. Each is accurately placed and named. About the same number of cities are named and no one mistake can be listed. Further, about thirty-six towns are mentioned, and most have been identified. Wherever accuracy can be checked, minute detail has been found correct-every time!" (1993, p. 129).

E. The Hittites are mentioned over forty times in the Old Testament (*e.g.,* Exod. 23:28; Josh. 1:4; 2 Kings 7:6). For centuries, critics scorned that the Hittites never existed. In spite of this, in 1906, Hugo Winckler discovered the Hittite capital along with 10,000 Hittite tablets. Hittite civilization is currently studied in university programs (Thompson, *op. cit.*).

Sir William M. Ramsay disbelieved the Bible until archaeological digs and research convinced him of the veracity of the Scripture (npgprints.com).

Human-headed winged bull or Lamassu at king Sargon II's palace at Dur Sharrukin in Assyria (now Khorsabad in Iraq) (wikicommons).

The Code of Hammurabi (discovered in 1901), written on this stele, represented human writing, which existed up to 300 years before Moses, disproving the idea that there was no written language in the day of Moses (wikicommons).

Dr. Hugo Winckler was a German historian and archaeologist who uncovered the Hittite capital in 1906, disproving the critics' contention that the Hittites were a fictional nation invented by Biblical writers.

Hittite god on a throne between two lions at Carchemesh. The Hittites were long scorned as Bible fiction until their archaeological discovery in 1906. (Photo from Hammerton, 1924, p. 561).

The Horites civilization, who critics said never existed, were discovered by archaeologists in 1925 along with their caves.

Archaeologists in 1961 discovered limestone with the Pontius Pilate's inscription in Caesarea, dispelling the critics who said Pilate never existed (wikicommons).

The Moabite Stone. Photo by Mike Willis.

One of the 382 tel-Amarna tablets, which were diplomatic letters between Egypt and foreign nations in Akkadian cuneiform writing. They confirm the custom in the Biblical account of bowing to the ground seven times in greeting and respect to one of higher standing during the time of Jacob (Genesis 33:3) (Photo: Wikicommons).

F. For decades, critics denied that Pontius Pilate, governor of Judea who sentenced Jesus to death, existed. However, in 1961 a limestone with his name engraved on it was discovered, showing the Bible to be historically accurate.

G. Genesis 33:3. "And he [Jacob] himself passed over before them, and bowed himself to the ground seven times, until he came near to his brother [Esau]." The custom of bowing to the ground seven times in greeting and respect to one of higher standing during the time of Jacob was verified by the Tel-el-Amarna Tablets (Thompson, *op. cit.*).

H. The Old Testament mentions the Horites (Gen. 14:6; 36:21). In 1925, archaeological digs proved the existence of these people (Free and Vos, 1992, p. 66). See photo above.

I. The Moabite Stone, Discovered in 1868. This table of writing was cut in 850 BC and recounts Mesha, king of Moab, becoming subject to the Israelites and Omri, captain of the Israelites, who was made king that day. This event is partially recorded in 1 Kings 16:16.

J. The Bible mentions King Belshazzar of Babylon (Dan. 5:22; 7:1; 8:1). Scoffers, again, sneered at the Bible for dreaming up such a fictional character as

Belshazzar, king of Babylon, until 1876 when Sir Henry Rawlinson discovered 2,000 Babylonian tablets. Guess whose name was mentioned as being a coregent with his father Nabonidus? Belshazzar, King of Babylon.

K. Attempting to discredit the Bible, critics said that Roman crucifixion was accomplished by tying the victims limbs and not nailing them to the cross. However, in 1968 a heel bone of a crucified man (Yehohanan) with a crucifixion nail traversing the heel bone was discovered in north east Jerusalem, disproving the critics.

III. Biblical Prophecy

Prophesying or predicting future events is an area where accuracy is beyond the pale of human possibility. "Predictive prophecy is the highest evidence of divine revelation. The one thing that mortal man cannot do is to know and report future events in the absence of a train of circumstances that naturally suggest certain

Cylinder of Nabonidus, which confirms the existence of his son King Belshazzar of Babylon, despite the critics previous protests (wikicommons).

Heel bone of crucified man discovered in north east Jerusalem in 1968 (Timesoflsrael.com).

possibilities... " (Turner, 1989, p. 12). The Bible says this much in stating that we should not believe one whose prophecies fail. Notice:

Deuteronomy 18:20-22

"But the prophet who presumes to speak a word in My name, which I have not commanded him to speak, or who speaks in the name of other gods, that prophet shall die. And if you say in your heart, 'How shall we know the word which the Lord has not spoken?'–when a prophet speaks in the name of the Lord if the thing does not happen or come to pass, that is the thing which the Lord has not spoken; the prophet has spoken it presumptuously; you shall not be afraid of him."

Prophets in the Bible and the Bible, itself, stake their credibility on their prophetic accuracy and, in that sense, Isaiah challenged the idolatrous prophets to do the same:

Isaiah 41:22

"Let them bring forth and show us what will happen; Let them show the former things, what they were, That we may consider them, And know the latter end of them; Or declare to us things to come."

Notice the following fulfilled biblical prophecies:

A. The destruction of Tyre (Ezek. 26:7-8). Nebuchadnezzar, king of Babylon was to destroy Tyre (26:7, 8). Multiple nations were to rise up against Tyre (26:3). The city would be leveled and scraped clean like a bare rock (26:4). Tyre's stones, timbers, and soil would be cast into the sea (26:12). Tyre would become a place where fishermen dried their nets (26:5). The city was never to be rebuilt to its original glory (26:14). This all happened just as predicted. It was destroyed by Nebuchadnezzar. Tyre was later attacked by Alexander the Great and the inhabitants

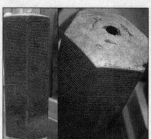

Sennacherib's prism, which states that, "he had shut up Hezekiah the Judahite within Jerusalem, his own royal city, like a caged bird" (wikicommons).

Relief of Cyrus II the Great, King of Persia who was predicted in the Bible, 150 years before he was born (Simax).

moved to an island off the coast while Alexander dismantled the city and made a land bridge to the island. Finally, Tyre was totally destroyed by Muslims in AD 1291. Today Tyre is a bald rock where fisherman spread and dry their nets (Bradshaw, 1999; Apologetics Press Staff, 1994, p. 96; Major, 1996, pp. 93-95).

B. God predicted He would use the Assyrians to punish Israel for their wickedness and then turn around and punish the Assyrians (Isa. 10:5, 6, 12, 24, 25). Abundant archaeological evidence confirms that this all occurred. It was also predicted that Sennacherib would return home to fall by the sword (2 Kings 19:7). Twenty years later, he was murdered by two sons while he was worshiping pagan idols (Isa. 37:37, 38).

C. Cyrus, King of Persia, the renowned historical figure, was foretold down to his name, 150 years before he was born (Isa. 44:28; 45:1).

Diagram of Alexander the Great's causeway—built with the ruins of Tyre cast into the sea as the Bible prophesied. (Courtesy of Apologetics Press.)

IV. The Scientific Accuracy of the Bible

Dr. Larry Dickens noted, "But the Bible does not flee from scientific information nor is it ever scientifically inaccurate. Just as statements of historical import are always accurate in the Bible, when the Bible does speak of some natural

Dr. Larry Dickens, former professor of chemistry, affirms the scientific accuracy of the Scripture any time natural science is addressed.

Dr. Doy Moyer, former professor of theology at Florida College, attributes Biblical scientific accuracy to divine inspiration.

Portuguese explorer, Ferdinand Magellan, in 1522 circumnavigated the earth demonstrating the world to be spherical. 2,500 years earlier, the Psalms affirmed that the earth was a KHUG, which can mean "sphere."

Photo of the World War II Liberty Ship, S.S. Jeremiah O'Brien. Liberty ship dimension ratios were similar to that of Noah's ark. (Photo from the National Liberty Ship Memorial Archives. Used with permission).

(scientific) truth, it is totally *and completely correct* (2004, p. 244). Doy Moyer stated, "There is scientific accuracy—far too much of it to be accounted for on the basis of lucky chance. The only way that these can be properly accounted for is by divine inspiration. Only God could know then what has taken men thousands of years to discover" (1995, p. 137).

This is not to say that other ancient cultures weren't knowledgeable of some scientific claims, only that the Bible (when addressing science) is never inaccurate.

A. Isaiah 40:22—"It is he Who sitteth upon the circle of the earth." The Hebrew word for circle is *khug* which can mean a sphere. In Isaiah's day, many thought the world was flat. Two thousand years later, Ferdinand Magellan's voyage round the world (AD 1519-1522) decisively demonstrated that the world is spherical. How did Isaiah know?

B. In approximately 2,000 BC, God told Noah (Gen. 6:15) to build an ark 300 x 50 x 30 cubits (or 450 feet long with 150 million cubic feet of space). These measurements produce a ratio of 30:5:3. The ark was the largest seagoing vessel ever built until 1858 when Isambard K. Brunnel launched the *Great Britain*. Guess what the ratio of its dimensions were? 30:5:3. These dimensions turn out to be scientifically precise for a ship designed to carry cargo and not for speed. In WWII, shipbuilders built the Liberty Ship or "Ugly Duckling" whose sole purpose was to carry cargo. Guess what the ratio of its dimensions were? Approximately 30:5:3. How did Noah know to build the ark with these precise dimensions hundreds of years before large ships even began to be built (Thompson, *op. cit.*)?

C. The Bible three times says that the earth is wearing out like a garment (Heb. 1:11; Isa. 51:6; Ps. 102:26). Not until recently was this principle proven. It is known as the second law of thermodynamics (the law of increasing entropy), which states that all things are becoming more disorderly, random, and chaotic. Everything is breaking down and deteriorating. Theoretically, given enough time, the universe itself will be exhausted of usable energy. How did biblical writers know this?

D. God commanded Abraham to circumcise newborn males on the eighth day after birth. We now know that at least three things are required for blood clotting: platelets, vitamin K, and prothrombin. On days 5-7 following birth, the body increases the basal amount of vitamin K by the probiotic action of symbiotic (healthy) bacteria in the intestinal tract. By day five, adequate amounts of vitamin K are made available for clotting, and by day eight, prothrombin in the infant is over 100% above normal levels. The eighth day is still known as the earliest day that surgery should be performed in infants. How did Abraham know this 4,000 years ago?

Conclusion

These are but a few of the innumerable examples of the historical infallibility, prophetic accuracy, and scientific accuracy displayed in the Bible. Time fails us to analyze the fallibility of religious documents such as the Koran, the Vedic Texts, writings of Buddha, the Book of Mormon, etc. and show even a smattering of their historical contradictions, archaeological falsehoods, scientific inaccuracies, and failed prophecies. The Bible is unlike any book ever written. It is your choice to believe the Bible as the Word of God or to reject it. For eons, skeptics who have attempted to discredit the Bible have inevitably failed.

1 Peter 1:23-25

"Having been born again, not of corruptible seed but incorruptible, through the word of God which lives and abides forever, because all flesh is as grass, and all the glory of man as the flower of the grass. The grass withers, and its flower falls away, but the word of the Lord endures forever. Now this is the word which by the gospel was preached to you."

Last eve I passed beside a
  blacksmith's door
And heard the anvil ring the vesper
  chime;
Then looking, I saw upon the floor,
Old hammers, worn out with the
  beating years of time.

"How many anvils have you had,"
  said I,
To wear and batter all these
  hammers so?" "Just one," said
  he, and then with twinkling eye;
"The Anvil wears the hammers out,
  you know."

And so, thought I, the anvil is God's Word,
For ages skeptic blows have beat upon;
Yet though the noise of falling blows was heard
The anvil is unharmed… the hammers are all
  gone.

—Author Unknown

(Photo: Wikicommons)

NOTES

_____

_____

_____

_____

_____

_____

Questions

MULTIPLE CHOICE

Circle all correct answers

1. The false concept that each person should be able to decide what morality is for themselves is called (a) Objectivity; (b) Tolerance; (c) Moral Relativism; (d) Islam.
2. How many cities does Luke mention in the book of Acts? (a) 13; (b) 5; (c) 20; (d) 54.
3. What biblically-mentioned civilization was at one time taught as being fictional? (a) Israelite; (b) Hittite; (c) Egyptian; (d) Horite.
4. What famous archaeological find mentions Omri, King of Israel? (a) The Ras Shamra tablets; (b) The Tel-el-Amarna tablets; (c) The city of Lachish; (d) The Moabite Stone.
5. The scientific law that confirms Psalm 102:26 (that the earth is wearing out like a garment) is (a) The law of relativity; (b) The second law of thermodynamics; (c) The law of biogenesis; (d) The law of gravity.

Short Answer

1. Name five major world religions and explain why all cannot be right. _____

_____

_____

_____

2. Explain why the unity of the Bible is such an incredible thing. _____

_____

_____

_____

3. How have historians proven that Moses could have written the Pentateuch? _____

_____

_____

_____

4. Why might prophecy be considered one of the most powerful evidences of the Bible's inspiration? _____

_____

_____

_____

5. What is the significance of the ark's dimension ratio of 30:5:3? _____

_____

_____

_____

Fill in the Blank

1. Without understanding _____ _____, one cannot properly understand man's origin.

2. The Bible is composed of _____ books written by _____ men over _____ years.

3. The biblical custom of _____ _____ _____ was confirmed by the Tel-el-Amarna tablets.

4. Among 2,000 Babylonian tablets discovered was the record of king _____ of _____ whose death is recorded in the book of Daniel.

5. Two Bible passages, _____ and _____ , tell us that a prophet who prophesies falsely is not from God. What does this teach us concerning religions with failed prophecies?

Discussion Question in Preparation for Answering Unbelievers and Critics

You are with a group of friends eating lunch and the subject of religion comes up. One person comments, *"You can't say that one religion is better than another, because they are all based on faith."* How do you respond? _____

_____

_____

_____

_____

_____

Lesson 3

Background to Darwinian Evolution: Variation within a Kind, and the Death of Darwinism

I. Evolution: Defining Terms

It might surprise you to learn that evolution actually does occur; however, the term evolution means different things based on the context in which it is used. The word "evolution" is of Latin origin from *evolvere* and means to unroll or change. The first definition Webster's Unabridged Dictionary lists for evolution is, "The act of unfolding or unrolling; hence, in the process of growth; development; as, the evolution of a flower from a bud, or an animal from the egg." Thus, any change is, technically, evolution, and is sometimes called "microevolution" when referring to natural observable changes within an animal kind. For example, there are over 1,300 varieties of roses; however, all roses could have come from a common ancestor at some point. Although roses can be bred to display different traits, all 1,300 varieties of roses are still roses. It is scientifically valid that this change could occur based on the *law of genetic variation*. This law teaches us that each successive generation of plants or animals will have slight, yet distinct, differences in their genes. Genes, which are discrete segments of a plant or animal's DNA, contain the biochemical instructions for each living organism—an encyclopedia of information. Differences that are apparent in humans due to genetic variation might be the color of eyes, hair, or skin, height, weight, or genetic flaws that could cause disease.

One of over 1,300 varieties of roses, which could have come from a single common ancestor and changed by genetic plasticity (Bridge).

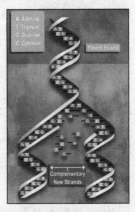
DNA carries the genetic blueprint for all living creatures. (Photo from www.ornl.gov/ hgmis).

Although there are hundreds of varieties of roses and thousands of varying traits in humankind, a rose is still a rose, and a human is still a human. Thus, we can easily witness genetic variation which is sometimes called *variation within a "kind,"* otherwise known as *microevolution*. That is to say, God created life forms in distinct groups, or kinds, with genetic boundaries that cannot be crossed. Notice the use of the word *kind* in the following verses:

Genesis 1:21

"So God created great sea creatures and every living thing that moves, with which the waters abounded, according to their kind, and every winged bird

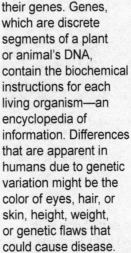
Humans vary in size. Sultan Kosen from Turkey, the tallest living man, stands 8'3" (Halldórsson).

Humans vary in height. Chandra Bahadur Dangi, the shortest man in the world from Nepal, stands 1'9" (Gwrthanesh).

Dogs, all descending from a common ancestor, vary in size, color, and disposition, all attributable to microevolution, or variation within a kind. But dogs never have and never will produce anything other than dogs, nor will they grow feathers instead of fur (Wilson).

A bird dog? The fixity of animal kinds prevents the evolution of one animal kind into another (Copyrighted photo. Used by permission).

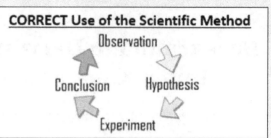

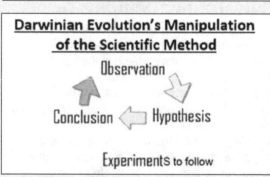

according to its kind. And God saw that it was good."

Genesis 1:24, 25

"Then God said, 'Let the earth bring forth the living creature according to its kind: cattle and creeping thing and beast of the earth, each according to its kind'; and it was so. And God made the beast of the earth according to its kind, cattle according to its kind, and everything that creeps on the earth according to its kind. And God saw that it was good."

You have probably observed the dog kind, the bird kind, the monkey kind, the bacteria kind, the fish kind, and the human kind. Within

Free Use Policy. © 2015 Answers in Genesis www.AnswersInGenesis.org

each kind there may be hundreds or thousands of genetic variations, varieties or types, based on microevolution or variation within a kind. However, a monkey will always be a monkey, a dog will always be a dog, and fish will always be a fish. Although we see changes in the variety of roses, which may all have a common ancestor, we never see a rose that turns into a sunflower. We may witness thousands of breeds of dogs that may have all descended from a common ancestor, but we never see a dog evolving into a cat. Further, there has never been any proof to support the hypothesis that the following succession of evolution occurred: bacterium --> amoeba --> red algae --> plankton --> protazoan --> fungi --> jellyfish and sponges -->trilobites and mollusks --> octopus and squids --> sea urchins and sand dollars --> plants --> fish and scorpions --> tetrapods --> insects and sharks --> crabs and amphibians --> reptiles and beetles --> ichthyosaurs --> dinosaurs and conifer trees --> birds --> flowering plants and bees --> hominids (great apes) --> snakes --> mammals: rodents --> primates --> whales --> bats --> camels --> felines --> sloths and dogs --> elephants --> mammoths --> hominids --> neanderthal --> humans.

Variation within a kind, or microevolution, is also referred to as the "special theory of evolution." This term should not be confused with the idea most commonly associated with the word "evolution": that of "organic evolution" or

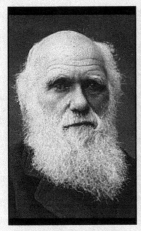

The last known photograph of Charles Darwin. (Photograph rights owned and administered by Henry E. Huntington Library and Art Gallery).

A sculpture of the Greek philospher Thales in 588 B.C. who proposed the idea of biological evolution more than 1200 years before Charles Darwin.

"macroevolution" or "neo-Darwinian evolution," commonly called "the general theory of evolution." (NOTE: See appendix A on the use of the terms "neo-Darwinian evolution" and the "modern synthesis" of evolution). Jon G. Williams, in his book, defines Darwinian evolution as, "The hypothesis that millions of years ago, lifeless matter acted upon by natural forces, gave origin to minute living organisms, which have since produced all extinct and living plants and animals, including man" (1996, p. 5). This is sometimes referred to as "particle to people," "monkey to man," or "atom to Adam" evolution. However, for simplicity we may refer to this doctrine as evolution or macroevolution, as it is known in the vernacular. Otherwise, the terms "Neo-Darwinism," "neo-Darwinian evolution," the "modern synthesis" of evolution, and the "modern evolutionary synthesis" will be used.

It may also surprise you to know that although it is taught that the doctrine of macroevolution is based on scientific fact; it is not. It cannot be. Remember the scientific method from Lesson 1? True science involves conducting reproducible experimentation in order to validate a hypothesis. When a hypothesis is tested multiple times and valid data attained, it is then possible to construct a theory. Because macroevolution supposedly took place over billions of years and is not experimentally reproducible, it is not even a theory, in the most scientific sense. It is only a hypothesis, which is simply an assumption, explanation, or an *idea*. Listen to Ernst Mayr, one of the architects of Neo-Darwinism in

his explanation that macroevolution is not scientifically testable. He said, "For example, Darwin introduced historicity into science. Evolutionary biology, in contrast with physics and chemistry, is a historical science—the evolutionist attempts to explain events and processes that have already taken place. Laws and experiments are inappropriate techniques for the explication of such events and processes. Instead one constructs a historical narrative, consisting of a tentative reconstruction of the particular scenario that led to the events one is trying to explain" (Mayr, 1999).

Renowned paleontologist Dr. John Horner and technical director for the movie *Jurassic Park,* said, "paleontology is a historical science, a science based on circumstantial evidence, after the fact. We can never reach hard and fast conclusions in our study of ancient plants and animals... These days it's easy to go through school for a good many years, sometimes even through college, without ever hearing that some sciences are historical or by nature inconclusive" (Horner, 1997).

For a recommended reading list of books and websites related to the Creation/Evolution discussion, see Appendix A2.

The National Geographic is no friend of Biblical creation.

II. The History behind Darwinian Evolution

Many believe that the modern idea of evolution began only after Charles Darwin's 1859 publication, *On the Origin of Species*. Many falsely concluded that evolutionary doctrine was some grand new budding scientific discovery, born out of the post-enlightenment era, the industrial revolution, and the age of reason during the past 150 years. Not so. Atheists and Greek philosophers have proposed this non-scientific hypothesis for centuries. Not until Darwin's writings, however, were any of these individuals taken seriously. Men such as Thales, 588 BC; Anaximander, 570 BC; Empedocles, 455 BC; Democritus, 420 BC; and Strato, 288 BC, all proposed that complex biological beings evolved from simpler ones and/or from non-living matter. Why would this idea be proposed? One reason is that up to 3,000 years ago "The fool has said in his heart, 'there is no God' " (Ps. 14:1; 53:1). Thus an alternative to divine creation was needed.

Even after Darwin's 1859 publication, no true scientific evidence was provided to support his macrospeculations. In fact, since Charles Darwin died, his ideas as to the mechanism of evolution have been thoroughly discarded by the scientific community as illegitimate. In the 1930's scientists incorporated the

creationist, Gregor Mendel's, gene theory with Darwin's macroevolution to come up with the "modern synthesis" of evolution, otherwise known as Neo-Darwinism. It is interesting that Darwin declared that animals evolved over millions of years leaving us with geological evidence of their gradual change from one transitional (i.e. intermediate) form to the next to the next. However, in almost the same breath, he freely admitted in his day that none of this fossilic evidence could be found. Notice his comments from *The Origin of Species*: "Why is not every geological formation and every stratum full of such intermediate links? Geology assuredly does not reveal any such finely-graduated organic chain; and this is the most obvious and serious objection which can be urged against the theory" (1872, pp. 264, 265). Although Darwin freely admitted that no transitional fossils had been found, he predicted that in time a more comprehensive survey of the geological column would uncover this evidence and validate his evolutionary hypothesis. To date, geological finds have, rather than fulfilling Darwin's prophecy, supported just the opposite. Fossils always appear in a fully developed form, rather than in the transitional, interphyla forms that Darwin predicted. This issue will be dealt with more extensively in Lesson 9.

III. Why Neo-Darwinian Evolution Has Been Accepted

Evolution might better be described as a religion rather than as a science since it is a hypothesis only, and is accepted on faith rather than by documented experiment-based conclusions. In reality, it takes more faith to believe that something came from nothing and that something intelligent came from something unintelligent, that evolved into the human race, than to believe that an intelligent Creator formed mankind by an act of will and intellect.

If you ask most people why they believe in evolution, their response will probably go something like, "I read it in the National Geographic," or "they teach it in schools and universities," or "I saw it on PBS and on The Discovery Channel," or "all educated people believe in evolution." In other words, most people believe in evolution because they have been *told they should* believe evolution. Kind of like the boy who believed in Santa Claus because "my daddy told me he came down the chimney." Don't believe me? Ask a few people and find out for yourself. Very few people, in truth, can even attempt to make a scientific case to support this hypothesis, much less present any factual evidence for it. Notice the following statement by Dr. Henry Morris, a reformed evolutionist turned creationist: "The writer is convinced, from having discussed the subject with hundreds of people, that the main reason most educated people believe in evolution is simply because they have been told that most educated people believe in evolution. Very rarely is such a person able to do more than repeat a few stock 'evidences for evolution,' and almost never has he given any really serious consideration to the question of their real implications" (1963, p. 26).

When some brave soul, however, steps out to challenge this scientific conjecture, he or she is scorned and derided, sometimes publicly, as being ignorant, uninformed, or an uneducated religious fundamentalist. I have seen it done. So, many in the scientific community, including students, although they have serious reservations and disagreements with the hypothesis of evolution, will pretend they believe the teaching for the sake of conformity. Dr. Thomas Dwight recognized this fact all the way back in 1911 in stating, "How very few of the leaders in the field of science dare to tell the truth as to the state of their own minds! How many feel themselves forced in public to do lip service to a cult that they do not believe in" (p. 21). Is this not reminiscent of Revelation 13:16, 17?

It is important to point out that nineteenth century Darwinian evolution arose, not from theists, but rather from atheists. Darwin himself gave up his belief in God prior to writing *On the Origin of Species* and later admitted that his book was an attempt to explain life on earth apart from God (Desmond and Moore,

1991, pp. 384, 386, 387; Thompson, 1999, pp. 49-55). Darwin gave up his religious faith at least twenty years prior to penning his famous work and concluded, "I can hardly see how anyone ought to wish Christianity to be true, for if so… men who do not believe, and this would include my father, brother and almost all my best friends, will be everlastingly punished. And this is a damnable doctrine" (Darwin, 1876, p. 87). Many of the atheistic leaders who originally embraced and proclaimed Darwinian evolution (such as Thomas Huxley) did so, not on the basis of sound scientific evidence, but rather because they were searching desperately for an alternative to divine creation. One professor at the University of Cincinnati said, "Our faith in the idea of evolution depends on our reluctance to accept the antagonistic doctrine of special creation" (More, 1925, p. 304). In fact, macroevolution, if not proved by empirical science, is simply a belief or… a **religion**. Notice what renowned evolutionist Dr. Michael Ruse admitted, "Evolution is promoted by its practitioners as more than mere science. Evolution is promulgated as an ideology, a secular religion—a full-fledged alternative to Christianity, with meaning and morality… the literalists are absolutely right. Evolution is a religion. This was true of evolution in the beginning, and it is true of evolution still today…Evolution therefore came into being as a kind of secular ideology, an explicit substitute for Christianity" (Ruse, 2000). Dr. James D. Bales in discussing the pro-evolution, anti-biblical bias observed, "If one is acquainted with the background of Darwin and other evolutionists in the nineteenth century, he will realize that they accepted evolution not because scientific evidence proved it, but because they had rejected the idea of creation by God and had determined that all must

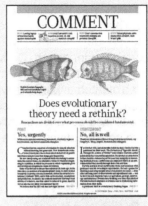

The journal Nature (2014) published a debate on whether the modern synthesis of evolution (Neo-Darwinism), which they call the "standard evolutionary theory" (SET) needs to be replaced with an "extended evolutionary synthesis" (EES).

be explained naturally" (Williams, 1996, p. 9). This anti-religious fervor extends to the present-day media as well. Earlier this year, in correspondence with Dr. Michael Behe, I asked him why he believes the mainstream media is so slanted toward Darwinism and against creationism. He responded, "I expect the media are biased because creationism is associated with conservatives" (Behe, 2005).

IV. "The League of Extraordinary Gentlemen" (Do Any Real Scientists Dispute Darwinism?)

Sometime back there was a movie with that title. I propose to you that there is an even greater league of extraordinary gentlemen and gentlewomen of the twentieth and twenty-first centuries who deserve accolades. These are those brave men and women in scientific, philosophical, and academic circles who have risen up to challenge the socially-accepted hypothesis of evolution. Their number is growing year by year. Some have long-held these ideas. Some were taught by others. And others were avowed evolutionists and atheists who studied the evidence on their own and came to the inevitable conclusion that the general theory of naturalistic evolution is flawed, despite their varying opinions regarding the earth's age. Further, some are not ready to abandon evolution altogether, they just want to make clear that neo-Darwinsm (as taught since the 1930's) is dead; thus, a new evolutionary mechanism is needed. In 2014, the journal *Nature* published a point-counterpoint debate on whether the modern synthesis of evolution (Neo-Darwinism), which they call the "standard evolutionary theory" (SET) needs to be replaced with an "extended evolutionary synthesis" (EES). SET supporters say that random mutations and natural selection are the key elements of macroevolution. Supporters of EES say that variation is not random and that a whole host of new evolutionary mechanisms play into the EES (*Nature*, 2014). To illustrate, the arguments taking place, EES proponents state, "Yet the mere mention of the EES often evokes an emotional, even hostile, reaction among evolutionary biologists. Too often, vital discussions descend into acrimony, with accusations of muddle or misrepresentation. Perhaps haunted by the spectre of intelligent design, evolutionary biologists wish to show a united front to those hostile to science" (*Nature*, 2014).

"A Scientific Dissent from Darwin"

In 2001, scientists signed *"A Scientific Dissent from Darwin,"* calling Neo-Darwinism into question. To be qualified to sign the dissent requires either a Ph.D. in a relevant scientific field or, if holding an M.D., one must also be a professor at an accredited university. There are currently over 950 signers. Signers include academics from all eight Ivy League schools, as well as from Stanford, Duke, Rice, Emory, Carnegie Mellon, Cal. Tech., Johns Hopkins, Vanderbilt, Stanford, U.C. Berkley, Univ. Chicago, Rutgers, MIT, UCLA, Tulane, McGill, Oxford, Cambridge, Manchester, Bristol, University of Nottingham, Imperial of London, Heidelberg University, Moscow State University, Seoul National University, Chitose of Japan, Ben-Gurion of Israel and the University of Tokyo. Source: http://www.dissentfromdarwin.org

There are, in fact, dozens of recently-published peer-reviewed scientific articles. A few of these have been provided in Appendix B: Selected Peer-reviewed Publications in Favor of Intelligent Design.

Dr. Eugene V. Koonin, National Center for Biotechnology Information (NCBI), National Institutes of Health (NIH): "The edifice of the modern synthesis [Neo-Darwinism] has crumbled, apparently beyond repair... So, not to mince words, the modern synthesis is gone." (Koonin, 2009).

Emeritus Professor David J. Depew and emeritus biochemistry professor, Bruce H. Weber, California State University, Fullerton: "Darwinism still rules the evolutionary roost... Darwinism in its current scientific incarnation has pretty much reached the end of its rope... We trace the history of the Modern Evolutionary Synthesis [Neo-Darwinism], and of genetic Darwinism generally, with a view to showing why, even in its current versions, it can no longer serve as a general framework for evolutionary theory" (Depew and Weber, 2012).

Former evolutionist, Dr. Henry Morris, critic of macroevolution, and founder of the Institute for Creation Research, Dallas, TX.

Dr. Henry M. Morris, former head of the Department of Civil Engineering, Virginia Polytechnic Institute: "Evolution doesn't happen, didn't happen and can't happen, and is fully unable to account for the design that we see" (2002).

Professor Colin Reeves, Dept of Mathematical Sciences, Coventry University: "What we have learned since the days of Darwin throws doubt on natural selection's ability to create complex biological systems – and we still have little more than handwaving as an argument in its favour" (Reeves, 2008).

Dr. Brian Goodwin (evolutionist and biologist, Open University, U.K). has been an ardent critic of the traditional view of evolution as taught in schools, and affirms that Neo-Darwinism can only demonstrate "microevolution," or, variation within a kind (Photo: Farnham, 1991).

Evolutionist Professor Brian Goodwin, biologist, Open University, UK: "Neo-Darwinism has failed as an evolutionary theory that can explain the origin of species, understood as organisms of distinctive form and behaviour. In other words, it is not an adequate theory of evolution. What it does provide is a partial theory of adaptation, or microevolution (small-scale adaptive changes in organisms)" (Goodwin, 1995b).

Evolutionist Professors Jerry Coyne, biologist, Univ. Chicago and H. Orr Allen, biologist, Univ. CA, Davis: "We conclude-unexpectedly-that there is little evidence for the neo-Darwinian view: its theoretical foundations and the experimental evidence supporting it are weak" (Allen & Coyne, 1992).

Chris Williams, Ph.D., Biochemistry Ohio State University: "Few people outside of genetics or biochemistry realize that evolutionists still can provide no substantive details at all about the origin of life, and particularly the origin of genetic information in the first self-replicating organism… Clearly the origin of life—the foundation of evolution – is still virtually all speculation, and little if no fact" (Williams, 2008).

Regarding the origin of life, Dr. Stanley Miller stated, "The problem of the origin of life has turned out to be much more difficult than I, and most other people, envisaged" (1991).

Dr. Michael Egnor, M.D. Pediatric surgeon and opponent of Neo-Darwinism.

Dr. Michael Egnor, Professor of Neurosurgery and Pediatrics at State University of New York, Stony Brook: "Darwinism is a trivial idea that has been elevated to the status of the scientific theory that governs modern biology… They've never asked scientifically, 'can random mutation and natural selection generate the information content in living things'" (*Dissent from Darwin*, 2015).

Dr. Ralph Seelke, professor of biology and neo-Darwinian opponent. Co-author of the textbook, "Explore Evolution: The Case For and Against Neo-Darwinism."

Dr. Ralph Seelke, microbiologist. professor in the Department of Biology and Earth Sciences at the University of Wisconsin-Superior: "Here's the point. If lots of soft-bodied animals existed before the Cambrian, then we should find lots of trace fossils. But we don't" (Dissent from Darwin, 2015).

Dr. Stanley Salthe, Renowned zoologist, and former Darwinist, Signer of "A Dissent from Darwin."

Dr. Stanley Salthe, Zoology, Professor Emeritus, Brooklyn College of the City University of New York: "Darwinian evolutionary theory was my field of specialization in biology. Among other things, I wrote a textbook on the subject thirty years ago. Meanwhile,

London Review of Books

Why Pigs Don't Have Wings

Jerry Fodor

Die Meistersinger is, by Wagner's standards, quite a cheerful opera. The action turns on comedy's staple, the marriage plot: get the hero and the heroine safely and truly wed with at least a presumption of happiness ever after. There are cross-currents and undercurrents that make *Meistersinger*'s libretto subtle in ways that the librettos of operas usually aren't. But for once Nietzsche is nowhere in sight and nobody dies, the territory is closer to *The Barber of Seville* than to *The Ring*. Yet, in the first scene of Act 3, the avuncular Hans Sachs, whose benevolent interventions smooth the lovers' course, delivers an aria of bitter reflection on the human condition. It comes as rather a shock:

> Madness, Madness!
> Madness everywhere.
> Wherever I look . . . .
> People torment and flay each other
> In useless, foolish anger
> Till they draw blood.
> Driven to flight,
> They think they are hunting
> They don't hear their own cry of pain . . . .
> When he digs into his own flesh,
> Each thinks he is giving himself pleasure.

So 'what got into Sachs?' is a well-known crux for Wagner fans, and one the opera doesn't resolve. (By Scene 2 of Act 3 Sachs is back on the job, arranging for Walther to get his Eva and vice versa.) Sachs isn't, of course, the first to wonder why we are so prone to making ourselves miserable, and the question continues to be pertinent. We have just seen the last of a terrible century with, quite possibly, worse to come. Why is it so hard for us to be good? Why is it so hard for us to be happy?

One thing, at least, has been pretty widely agreed: we can't expect much help from science. Science is about facts, not norms; it might tell us how we are, but it couldn't tell us what is

Dr. Jerry Fodor rocked the evolutionary world with his 2007 article, "Why Pigs Don't Have Wings, in which he proclaims the inability of Neo-Darwinism to account for macroevolution (Fodor, 2007).

however, I have become an apostate from Darwinian theory and have described it as part of modernism's origination myth. Consequently, I certainly agree that biology students at least should have the opportunity to learn about the flaws and limits of Darwin's theory while they are learning about the theory's strongest claims" (*Dissent from Darwin*, 2015).

Evolutionist Dr. Steven J. Gould, Professor of Paleontology, Harvard University: "Around 1980, Gould moved on to a more radical position, which has been actively discussed ever since. He turned the theory of punctuated equilibrium back on the modern synthesis [i.e., Neo-Darwinism] and (he claimed) toppled that synthesis off its throne. In 1980, he described the synthetic theory of evolution (Neo-Darwinism) as 'effectively dead'" (Ridley, 1996, p. 570).

In the book *What Darwin Got Wrong,* two evolutionists admit, "In fact, we don't know very well how evolution works. Nor did Darwin, and nor (as far as we can tell) does anybody else. 'Further research is required', as the saying

Three years later (in 2010) Dr. Jerry Fodor teamed with Dr. Massimo Piattelli-Palmarini (professor, University of Arizona) in publishing "What Darwin Got Wrong" again making a strong case against the ability of Neo-Darwinism to account for macroevolution (Fodor and Piattelli-Palmarini, 2010). (Photo: Courtesy of Salon.com) NOTE: See more quotes from this book in Appendix H: "More quotes from mainstream evolutionists who say that Neo-Darwinism is dead."

goes. It may well be that centuries of further research are required... we think that what is needed is to cut the tree at its roots: to show that Darwin's theory of natural selection is fatally flawed" (Fodor and Piattelli-Palmarini, 2010, p. xiv).

Dr. Michael Behe, Professor of Biochemistry, Lehigh University: "From Mivart to Margulis, there have always been well-informed, respected scientists who have found Darwinism to be inadequate." And, "Darwinism is an inadequate framework for understanding the origin of complex biochemical systems" (Behe, 1996, pp. 30,176).

Professor Scott F. Gilbert, Department of Biology, Swarthmore College; Dr. John M. Opitz, Foundation for Developmental and Medical Genetics; Professor Rudolf A. Raff, Department of Biology, Indiana University: "The Modern Synthesis is a remarkable achievement. However, starting in the 1970s, many biologists began questioning its adequacy in explaining evolution. Genetics might be adequate for explaining microevolution, but microevolutionary changes in gene frequency were not seen as able to turn a reptile into a mammal or to convert a fish into an amphibian. Microevolution looks at adaptations that concern only the survival of the fittest, not the arrival of the fittest" (Gilbert *et al.*, 1996)

Dr. Philip Skell. emeritus, Evan Pugh Professor of Chemistry, Pennsylvania State University Member of the National Academy of Sciences, father of carbene chemistry

and the "Skell Rule." "My own research with antibiotics during World War II received no guidance from insights provided by Darwinian evolution. Nor did Alexander Fleming's discovery of bacterial inhibition by penicillin. I recently asked more than 70 eminent researchers if they would have done their work differently if they had thought Darwin's theory was wrong. The responses were all the same: No" (*Dissent from Darwin*, 2015).

Dr. Henry Schaefer, 5-time Nobel Prize nominee and signer of "A Dissent from Darwin."

Professor Henry Schaefer, Director, Center for Computational Quantum Chemistry University of Georgia, five time Nobel Prize nominee: "There is no plausible scientific mechanism for the origin of life, i.e., the appearance of the first self-replicating biochemical system... I find no satisfactory mechanism for macroevolutionary changes" (*Dissent from Darwin*, 2015).

Dr. Egbert Leigh, Smithsonian Tropical Research Institute Staff Scientist Emeritus, Evolutionary Biologist, Ecologist: "The 'modern evolutionary synthesis' convinced most biologists that natural selection was the only directive influence on adaptive evolution. Today, however, dissatisfaction with the synthesis is widespread, and creationists and antidarwinians are multiplying. The central problem with the synthesis is its failure to show (or to provide distinct signs) that natural selection of random mutations could account for observed levels of adaptation" (Leigh 1999).

Lynn Margulis, evolutionist, American biologist and university professor in the Department of Geosciences at the University of Massachusetts Amherst; 2008 Darwin-Wallace Medalist; ex-wife of the late Carl Sagan: "At that meeting, Ayala agreed with

Professor Lynn Margulis, evolutionist, Univ. of MA, Amherst. Denies that Neo-Darwinism is capable of producing macroevolution (Jpedreira).

me when I stated that this doctrinaire NEO-DARWINISM IS DEAD. He was a practitioner of Neo-Darwinism but advances in molecular genetics, evolution, ecology, biochemistry, and other news had led him to agree that Neo-Darwinism is dead" (Margulis, 2010, p. 285).

"Neo-Darwinian language and conceptual structure itself ensures scientific failure: Major questions posed by zoologists cannot be answered from inside the neo-Darwinian straitjacket. Such questions include, for example, 'How do new structures arise in evolution?' 'Why, given so much environmental change, is stasis so prevalent in evolution as seen in the fossil record?' 'How did one group of organisms or set of macromolecules evolve from another?' The importance of these questions is not at issue; it is just that neo-Darwinians, restricted by their presuppositions, cannot answer them" (Margulis and Sagan, 1997).

Other respected scholars who have rejected neo-Darwinian evolution and signed the *Dissent from Darwin* petition include the following individuals, all with doctoral degrees in their respective fields: Prof. Maciej Giertych (Polish Acad. of Sci.), Prof. Lev Beloussov (Russian Acad. of Sci., Moscow State Univ.), Dr. Eugene Buff (Russian Acad. of Sci.), Ferenc Jeszenszky (Hungarian Acad. of Sci.), Denis Fesenko (Russian Acad. of Sci.), David Chapman (Senior Sci., Duke Univ.), Prof. Marcos N. Eberlin (Brazilian Acad. of Sci.), Prof. John C. Walton (Reactive Chem., Royal Society of Chem.), Leo Zacharski (Prof. of Med., Dartmouth Univ.), Michael J. Kavaya (Senior Scientist, NASA Langley Research Ctr.), Richard Spencer (Prof., U.C. Davis, Solid-State Circuits Research Lab.), Richard Austin (Prof. & Chair, Biology & Natural Scie. Piedmont College), William J. Arion) Emeritus Prof. of Biochem. Cornell Univ.), Mark L. Psiaki (Prof. of Mechan.l and Aerospace Engineering (Cornell University), Russel Peak (Senior Researcher, Engineering Info. Systems, Georgia Tech.), Joe R. Eagleman (Prof. Emeritus, Dept. Physics & Astron., Univ. of Kansas), Dewey Hodges (Prof., Aerospace Engineering, Georgia Tech.), Nolan Hertel (Prof., Nuclear & Radiological Engineering, Georgia Tech), Graham Gutsche (Emeritus Prof. Physics, U.S. Naval

Acad.), David Johnson (Prof. Pharmacol. & Toxicol., Duquesne Univ.), Rodney Ice (Principle Research Sci., Nuclear & Radiolog. Engineering, Georgia Tech.), Raleigh R. White IV (Prof. of Surgery, Texas A&M Univ., College of Med.), Jay Hollman (Prof. Cardiology, LSU, Health Sci. Ctr), Scott T. Dreher (Geology, Royal Society, USA Research Fellow, Univ. of Alaska, Fairbanks), Edwin Karlow (Chair, Dept. Physics, LaSierra Univ.), P.G. Ackerman (Prof. of Psych., Wichita St. Univ.), J.R. Baumgardner (Geophysicist, Univ. of Cal., Los Alamos Natl. Lab.), R.W. Carlson (Prof. Biochem. and Molec. Biol., Univ. of Ga.), P. Chien (Prof. Biol., Univ. of San Francisco), V. Damadian, M.D. (Nobel Nominee and inventor of the MRI), D.K. DeWolf (Prof. Law, Gonzaga School of Law), D. Faulkner (Prof. Astronomy, Univ. of S. Carolina, Lancaster), C.B. Fliermans (Dupont Microbiologist and discoverer of Legionnaire's Disease), G. Gonzalez (Astrophysicist and Prof. Astronomy, Iowa St. Univ.), D. Hamann (Food Eng., Former Prof., Dept. Food Sci., N. Carolina St. Univ.), W.S. Harris, Ph.D. (Biochemist and Prof. Med., Univ. of Missouri, Kansas City), R. Hirsch (Structural Biol., U.S. Dept. of Energy), C. Johanson (Prof. Clinical Neurosci., Brown Univ.), R. Kaita (Principal Research Physicist, Plasma Research Lab., Princeton, Univ.), D.H. Kenyon (Prof. Emerit. Biol., San Francisco St. Univ.), R.C. Koons (Prof. Philos., Univ. of Tex.), S. Lennard, M.D., Sc.D. (Prof. Surg., Univ. of Wash. School of Med.), S. Minnich (Prof. Microbiol., Univ. of Id.), H.K. Ortmeyer (Prof. Physiol., Univ. of Md.), J.M. Schwartz, M.D. (Prof. Psych., UCLA School of Med.), Hyun-Kil Shin (Prof. Food Chem., KunKuk Univ., Korea), F. Skiff (Prof. Physics, Univ. of Iowa), F. Tipler (Prof. Mathematical Physics, Tulane Univ.), and R. Weikart (Prof. Hist., Cal. St. Univ.).

To make more clear the astronomical improbability of random mutations and natural selection (Neo-Darwinism) producing *JUST ONE FUNCTIONAL PROTEIN* of at least 150 amino-acid sequences, Dr. Douglas Axe (2004) published the results of his experiment to produce such a protein. In the prestigious *Journal of Molecular Biology,* he concluded that producing *ONE FUNCTIONAL PROTEIN* by random mutations would take 10^{77} (a number equivalent to a 1 with 77 zeros after

it) iterations (or tries). That is, one chance in one hundred thousand, trillion, trillion, trillion, trillion, trillion, trillion. To put that in perspective, scientists estimate that in the history of the world (which they think is 4.5 billion years old) there have only be 10^{40} animals that have ever lived (Meyer, 2013, p. 203). Thus, the chances of producing *JUST ONE FUNCTIONAL PROTEIN* go from improbable, to impossible.

Dr. Douglas Axe, whose 2004 publication demonstrated that the chances of producing one functional protein are one in 10^{77} tries. Written out, this is one chance in 100000000000000000 000000000000000000000000 000000000000000000000000 0000000000000 tries. However, there have only been 10^{40} animals that have ever lived.

Yet, evolutionists claim that these functional proteins have been created by mutations billions and billions of times. Now it becomes more clear why Neo-Darwinism must be rejected.

Conclusion

If Neo-Darwinism is dead, then is macroevolution entirely rejected? Unfortunately, the answer to that question is, no. Remember, these are men and women of great faith. They BELIEVE that macroevolution MUST be true, even with no mechanism as to how it occurred. For example, evolutionist Barry Price said, "While evolution is a fact, how it occurs will always be the subject of debate. This is the fascination of science. To put it another way, there is no dispute about the fact that evolution has occurred, but there is dispute among scientists about how it has occurred" (1990). From the failure of Neo-Darwinism has arisen a multitude of competing evolutionary hypotheses; however, no one can agree on which mechanism to accept, because most of them seem even harder to believe than Neo-Darwinism. Some of the more common ND alternatives are as follows: Evolutionary Development (i.e., Evo-devo), Genetic flow, Genetic hitchhiking, Genetic drift, Biased Mutation, Directed mutation (adaptive evolution) (Cairns), Epigenetic Inheritance, Niche Construction, Self-organization (Kauffman *et al.*), Whole Genome Doubling (Shapiro), Orthogenesis (Directed Evolution) (Lima-de-Faria), Salatationism (Balon,

AFTER EDEN   by Dan Lietha

PARENTS, DON'T TEACH YOUR KIDS TO BE BIBLICAL CREATIONISTS. WE NEED THEM TO BELIEVE IN EVOLUTION SO THEY CAN BECOME COMPETENT SCIENTISTS AND ENGINEERS FOR THE FUTURE.

"Yeah, I wouldn't want to grow up to be an incompetent scientist like Isaac Newton, Francis Bacon, Louis Pasteur, Johann Kepler, Wernher von Braun, Michael Faraday, . . ."

Free Use Policy. © 2015 Answers in Genesis
www.AnswersInGenesis.org

Norrstrom), Process Structuralism, Symbiotic reorganization (i.e., symbiogenesis) (Margulis *et al.*), Punctuated equilibrium, Natural genetic engineering (Shapiro), Neo-Lamarckism (Jablonka, Pigliucci), Facilitated variation (Gerhart and Kirschner), The Gaia hypothesis, Teleonomic Selection (Corning), Holistic selection, Nearly neutral theory (Moran *et al.*), Social selection (Roughgarden), Environmental fitness (i.e., physical/biochemical fine-tuning), Intrinsic biotic constraints (i.e., Directionality), Group selection (E.O. Wilson). Until a consensus can be reached on an "Extended Synthesis" to ND, evolutionists would like the general public to be in the dark regarding the death of ND and the bitter disagreements over what new evolutionary mechanism to accept (NOTE: See Appendix C for a list of mainstream evolutionary publications calling for alternatives to ND).

Now you know what macroevolution is, its origin, and why it has been accepted by some while being rejected by others. In subsequent lessons we will put various aspects of the hypothesis to the test.

NOTES

_____
_____
_____
_____
_____
_____

Questions
TRUE OR FALSE

1. ___ The word *evolution* comes from a Latin word that means to change from one kind of animal into another.

2. ___ All 1,300 varieties of roses could have come from a common ancestor.

3. ___ If one believes in microevolution, then he or she must also believe in macroevolution.

4. ___ Scientists today accept Darwin's evolutionary mechanism while discarding his general theory of evolution.

5. ___ The transitional fossils that Darwin predicted would be found in the geological strata have been uncovered.

6. ___ No true scientist renounces the general theory of evolution.

Fill in the Blank

1. _____ contains biochemical instructions for living organisms that allows for variation within a kind.

2. The Bible passage _____ teaches us that God created all animals according to their own kind.

3. Charles Darwin's book, much of which has been proven fallacious, where he first stated his theory of evolution is called _____.

4. Charles Darwin gave up his belief in _____ prior to publishing his evolutionary agenda.

5. Other names for atheistic evolution are _____ evolution, _____ evolution, _____ evolution or _____ _____ _____ ___ evolution.

6. One of the earliest supporters of evolution was _____ who lived in the year _____.

Short Answer

1. Explain the difference between micro- and macroevolution. _____
   _____
   _____

2. What are transitional fossils and why is finding them important to the evolutionist? _____
   _____
   _____

3. Why do you suppose Charles Darwin's theory became so popular in the nineteenth century? Do you believe it became prominent then for the same reason it is popular today? _____
   _____
   _____

4. Why is the theory of evolution not a true theory in the strictest scientific sense? _____

_____

5. Why could evolution best be described as a religion? _____

_____

_____

6. What leads most people to accept the evolutionary hypothesis? _____

_____

_____

Discussion Question in Preparation for Answering Unbelievers and Critics

One day your biology teacher or science teacher asks for a show of hands of all those who believe in evolution. Only a few students raise their hands. She then shows you a diagram delineating how all dogs "evolved" from a common ancestor and says, *"If you can understand this diagram, then you must believe in evolution."* A second time she asks to see a show of hands of all those who believe in evolution. This time everyone in the class raises their hand except for you. She then calls you by name and asks, *"Don't you believe that it was possible for all these varieties of dogs to come from a common ancestor?"* to which you reply, *"Yes, probably."* She then asks you very pointedly, *"Then how can you say you don't believe in evolution?!"* How do you respond?_____

_____

_____

_____

_____

_____

_____

_____

Lesson 4

The Failings of Theistic Evolution (I): Inorganic (Non-Living) Evolution and the Days of Creation

Introduction

The first and second chapters of Genesis recount God's creation of the world in six, consecutive twenty-four-hour solar days. Christians and discerning Bible students understand and accept God's clear teaching on this subject. Even some non-believers who are students of the Bible readily admit that Genesis one and two records that the Bible teaches that the universe was created in six twenty-four-hour days. Notice the following quote by Dr. James Barr, Regius professor of Hebrew at the prestigious Oxford University who did not accept the inerrancy of the Bible: "So far as I know, there is no professor of Hebrew or Old Testament at any world-class university who does not believe that the writer(s) of Gen. 1-11 intended to convey to their readers the ideas that (a) creation took place in a series of six days which were the same as the days of twenty-four-hours we now experience" (as quoted in Grigg, 1993). There is, however, an alternative idea to neo-Darwinian evolution or Divine creation. This middle ground is commonly called theistic evolution.

I. What is Theistic Evolution?

Theistic evolution goes by many names including progressive creationism, religious evolution, mitigated evolution, old-earth creationism, and threshold evolution. Theistic evolution is the compromising position whereby one does not embrace a naturalistic (atheistic or humanistic) evolutionary belief, but at the same time rejects a literal interpretation of the creation week recorded in the first two chapters of Genesis. Theistic evolutionists assert that God either allowed or guided the evolutionary processes that Darwinism espouses. Those who walk this fine line of faith often do so in order to retain allegiance to God while attempting to remain popular with the humanistic scientific community. The entertainment industry, mass media, and even public schools and universities have made popular the myth that all credible scientists and scholars accept evolution as fact. Thus, many believers have been coerced into thinking they too must accept some form of this false teaching in order to be considered scientifically and socially relevant. "One important reason, however, for the current popularity of this false doctrine is that Christians have become intimidated. They have been intimidated by fancy 'facts,' impressive credentials, and flowery words and phrases set forth by some in

"In Six Days" contains 50 essays by scientists from diverse fields who argue for a literal six-day creation week as described in Genesis 1, along with scientific rebuttals to frequent objections to the "young earth" view.

Dr. James Barr (Oxford University) who did not accept the inerrancy of the Bible, nevertheless, admitted that the Bible taught that God created the world in six literal solar days.

In "Coming to Grips with Genesis: Biblical Authority and the Age of the Earth," fourteen biblical scholars with advanced degrees in theology exegete the Scripture in a detailed fashion, argue for a literal six-day creation week from Genesis chapter one. They affirm that a reasoned interpretation of the Bible must be the final authority over current inaccurate scientific dogma.

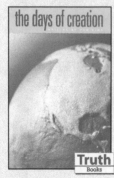

Gospel evangelist and author, Dr. Dan King Sr., in the book, "The Days of Creation" argues for the Biblical teaching of the six days of creation as taught in Genesis chapter one. This book is available through CEI Bookstore or other Bible bookstores.

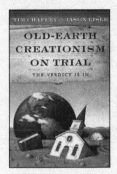

The Creation Wiki is a free educational encyclopedia that has been written from the creationist worldview (www.creationwiki.org).

the scientific community. They have become victims of propaganda campaigns which teach that 'anyone entitled to a judgment' believes in evolution, that 'all reputable scientists' accept evolution, that 'belief in evolution is a part of the learning process in all thinking persons,' and so on" (Thompson and Jackson, 1992b, p. 91).

The truth is that there are multitudes of scientists around the world who reject the evolutionary premise of Neo-Darwinism. Some of these individuals are members of various organizations such as the *Access Research Network, Answers in Genesis, Apologetics Press, Creation Research Society, Creation Science Foundation, Institute for Creation Research,* and *Intelligent Design Network.* What we must never forget as Christians is that regardless of what the so-called scientists, experts, and authorities avow, God's word is still the final and sole authority regarding our origin. Scientific theories are in a constant state of flux, undergoing modification, and even frequently disproved and discarded. God's word never changes, God's word should never be modified, and God's word can never be disproven. You cannot always rely upon secular science. You can always depend upon the word of God. "*Indeed, let God be true but every man a liar*" (Rom. 3:4). "When you received the word of God which you heard from us, you welcomed it not as the word of men, but as it is in truth, the word of God" (1 Thess. 2:13). "Every word of God is pure.... Do not add to His words, lest He rebuke you, and you be found a liar" (Prov. 30:5-6).

II. What Is Inorganic Evolution?

All known life forms derive their energy from and are composed of the element carbon, which makes up *organic matter*. The word "organic," thus, is defined as, "of, relating to, or derived from living organisms." Darwinism typically deals with this type of evolution; that is, the molecule -to-man evolution of living creatures. However, many who would deny that teaching sometimes compromise the literal account of creation in the book of Genesis in another way. They reject the myth that organic living beings evolved from atom to Adam over billions of years, yet they accept the doctrine of *inorganic* evolution—the belief that the earth, our solar system, and our universe evolved over untold *billions* of years prior to life being placed on this planet. In fact, the latest figure accepted by evolutionists for the age of the earth is approximately 4.6 *billion* years, and 13.8 billion years for the universe. Although the Genesis account clearly teaches what we might call a "creation week," because of secular teaching, many compromising Bible students are willing to discard the literal truth of the first and second chapters of the Bible in order to be in agreement with secular scientific dogma. Notice the following quote by one compromising Bible expert published in a book by the esteemed denominational Moody Bible Institute regarding a literal six-day creation: "This seems to run counter to modern scientific research, which indicates that the planet earth was created several billion years ago" (Archer,

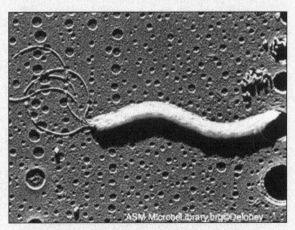

Micrograph of the bacterium Helicobacter pylori, proven to be the source of gastric ulcers in 1983. Science had previously attributed stomach ulcers to stress. Science is frequently disproven. The Bible is not. (Copyright C. DeLoney-Marino. Licensed for use, ASM Microbe Library).

1994, pp. 196, 197). Thus, more authority and credibility is given to man's hypotheses than to God's infallible intellect. "It is as if these theologians view 'nature' as a '67th book of the Bible,' albeit with more authority than the 66 written books" (Batten, 2003, p. 35).

Some Christians have even embraced the idea that the Bible should be interpreted by science rather than science by the Bible. If one simply considers many of the wrong ideas science has embraced over the years, it will become abundantly clear that fallible science should always be subject to God's holy and inerrant word, not the other way around. For instance, in the past, the secular scientific community held such faulty views as a flat earth. They taught a geocentric instead of a heliocentric solar system, that is the idea that the earth is the center of the solar system instead of the sun. Physicians at one time believed that bad blood was responsible for individuals' illnesses. In attempts to heal people, these doctors drained pints of the patient's life blood by "bleeding" them out, killing many people, including the first president of this country. What about the scientific evidence that prescription drugs are safe, despite thousands of yearly deaths from prescription drugs in the U.S. Some drugs that have been implicated include:

1976-1997: The FDA-approved drug Fen-Phen led to Heart valve destruction

1962-2008: Fluoroquinolone drugs led to Ruptured tendons

1999-2004: Vioxx induced 90,000-144,000 cases of heart disease

1955-2010: Darvocet led to heart arrhythmias (this is the most commonly-prescribed euthanasia drug)

2000-2010: Avandia caused 83,000 heart attacks in 8 years!

Other previously dispelled scientific ideas include the teaching that all life forms do not depend on solar energy. Science said the sun revolved around the earth, science said

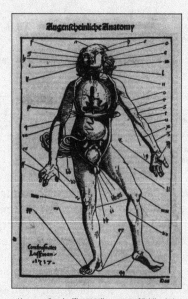

Hans von Gersdorff's 1517 illustration of Feldbuck der Wundarznei or "Points for blood-letting."

that fire is caused by the release of a substance called phlogiston, Charles Darwin said that inheritance was transmitted through "gemules" in the body by a process called "Pangenesis," lobotomies were used in the 1940's to treat the mentally ill, and the Steady State theory of the universe was replaced by the "Big Bang" theory in 1964. For over a decade, science said that autism was caused by childhood vaccines, until this fraudulent publication was retracted in 2010. Science taught that a living organism can spontaneously appear from a soupy broth, that the atom could not be split, and that the African races evolved as a species intellectually inferior to Europeans. This last idea was trumpeted by Charles Darwin, himself. Whereas all of these ideas have been proven wrong, God's word has not been proven wrong and has not changed. And now we are told we must interpret the Bible according to modern secular scientific thought? Who can believe it? "The grass withers, And its flower falls away, But the word of the Lord endures forever" (1 Pet. 1:24-25). "It is amazing that men will accept long, complicated imaginative theories and reject the truth given to Moses by the Creator Himself" (Riegle, 1962, p. 24). It is a pity that those who claim to be a "guide to the blind, a light to those who are in darkness, an instructor of the foolish, a teacher of babes" (Rom. 2:19-20), are not satisfied to simply, "speak as the oracles of God" (1 Pet. 4:11).

Dr. Terry Mortenson wrote an article citing 12 renowned Biblical scholars who say they believe the earth is 4.5 billion years old, not because the Biblical text demands it, but because they have allowed secular science to take precedence and have the final authority over the plain teaching of Genesis chapter one. Dr. Mortenson states, "The Christian scholars cited above and many other evangelical scholars and leaders during the past 200 years all say basically the same thing in different words. In essence, they are teaching

the church that science is the final authority in determining the correct interpretation of some or all of Genesis 1–11, or at least that science is the final authority in determining that the young-earth view must be wrong. Therefore, they think, Bible scholars are free to advocate all kinds of alternative interpretations, no matter how exegetically weak they may be" (Mortenson, 2010). As an example, Dr. Mortenson quotes J.P. Moreland who says, "The date of creation is a difficult question, but on exegetical grounds alone, the literal twenty-four-hour-day view is better. However, since the different progressive creationist views are plausible exegetical options on hermeneutical grounds alone, then if science seems to point to a universe of several billions of years, it seems allowable to read Genesis in this light" (Moreland, 1998). So, Dr. Moreland concludes that the interpretation of Genesis must be subservient to secular science.

One of the most common tenets of inorganic theistic evolution is that each day in the six days of the Genesis creation account was not a literal twenty-four-hour day but was a vast stretch of time composed of thousands, millions, or billions of years. This is commonly called the day-age hypothesis. In fact, some Christians even teach the doctrine that all six days of creation put together comprised over 13.8 *billion* years, the secular-accepted age

of the universe. Let us now examine why this hypothesis is wrong.

III. Why Inorganic Evolution (or the Day-age View) Must Be Rejected

Although it is sometimes tempting to fall in with the crowd to gain their approval, sincere service to our Lord and Savior demands that we must adhere to the truth of His word. "The entrance of Your words gives light; it gives understanding to the simple" (Ps. 119:130). The following are a few of the many reasons why we should believe that inorganic evolution and the day-age theory are wrong, that the earth did not evolve for billions of years prior to man's introduction, and that the word *day* in the *Genesis* creation account refers to a normal twenty-four-hour day and not extended ages of time.

A. *Genesis* chapters one and two state that God created the world in *six days.*

B. The word "day" in *Genesis* one and two is from the Hebrew word *yom,* which is used 1284 times in the Old Testament. Although the word, in rare instances, can refer to a period of time (*e.g., during the day of Abraham Lincoln"*) this fact is clearly borne out in the context of the passage and typically is not preceded by a numerical reference (see Gen. 2:4; Ps. 95:8, 9; and Jer. 46:10). However, whenever *yom* follows a numeral in non-prophetic writings in the Old Testament (such as in Genesis) it always has reference to a twenty-four-hour

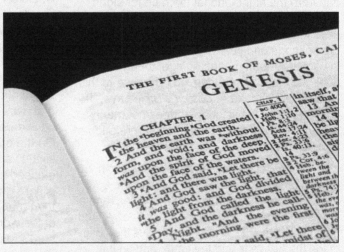

A literal interpretation of the first two chapters of Genesis is under attack by some religious thinkers. (Photograph rights purchased from istock.com).

"The Bible and the Age of the Earth," which argues for a literal six-day creation week, also responds to the most popular theological arguments for an "old-earth." This 149-page book is available for purchase or FREE in PDF format from www.apologeticspress.org.

solar day. One biblical scholar noted, "We have failed to find a single example of the use of the word 'day' in the entire Scripture where it means other than a period of twenty-four hours when modified by the use of the numerical adjective" (Williams, 1965, p. 10).

C. One survey of Hebrew scholars in nine prominent universities conducted by a Canadian anthropologist corroborated the Biblical twenty-four-hour day position. The professors were asked, "Do you understand the Hebrew yom, as used in Genesis one, accompanied by a numeral, to be properly translated as (a) a day as commonly understood, or (b) an age, or (c) an age or a day without preference for either?" Of the seven of nine Hebrew scholars that responded to the survey, all affirmed that *yom* in Genesis one was referring to a normal twenty-four-hour solar day (as quoted in Surburg, 1959, p. 61).

D. With regard to the word "day" being an age of incalculable length, notice Genesis 1:14: "Then God said, 'Let there be lights in the firmament of the heavens to divide the day from the night; and let them be for signs and seasons, and for days and years.'" If the days in Genesis one are ages, then what are the seasons and years? Longer ages? If *day* in verse fourteen means an age, then what does the word *night* mean? In reference to this, Marcus Dods in the *Expositor's Bible* says, "If the word 'day' in this chapter does not mean a period of twenty-four-hours, the interpretation of Scripture is hopeless" (1948, pp. 4-5).

E. Moses taught these were twenty-four-hour days in Exodus 20:11: "For in six days the Lord made the heavens and the earth, the sea, and all that is in them, and rested the seventh day." And again in Exodus 31:17, "for in six days the Lord made the heavens and the earth, and on the seventh day He rested and was refreshed." The word Moses used for days is the Hebrew word *yamim. Yamim* appears over 700 times in the Old Testament and in each instance in non-prophetic literature (such as in Genesis) it always carries the meaning of a twenty-four-hour period.

F. The days in Genesis one and two should be understood as normal twenty-four-hour days because they are accompanied by the phrase, "morning and evening" in Genesis one verses 5, 8, 13, 19, 23, and 31. Apologist Dr. Henry M. Morris stated, "The Hebrew words for 'evening' and 'morning' occur over 100 times each in the Old Testament and always in a literal sense" (Morris, 1970, p. 58, emphasis in the original).

G. How could God have made the fact that there were six twenty-four-hour days of creation any clearer? What else could God have said? "Six days and on the seventh He rested." "Morning and evening." "The second day." "The third day." "The fourth day," and so on. He was more than clear for any honest soul.

H. If the Holy Spirit, through Moses, had intended to mean ages instead of twenty-four-hour days in Genesis one, He could have employed one of the Hebrew terms for long periods of time: *olam* or *qedem.*

I. Consider this. If this planet was allowed to evolve for 4.6 *billion* years after which God created mankind, man would have been created at the *end* of creation wouldn't he? Imagine that all 4.6 billion years of the assumed evolutionary time were represented by one sixty-minute hour. In this illustration, animals would only have appeared in the last ten minutes, while humans would have only arrived on the scene in the last 1/100 second. Our Lord and Savior Jesus counters this idea in Mark 10:6 by saying, "But from the beginning of the creation, God made them male and female." But, if the earth is 4.6 *billion* years old, then man was created 4.5 *billion* years too late to be considered present at the beginning of creation. On the other hand, if man was created on the sixth twenty-four-hour day of creation, and the elapsed time from Adam till the present-day is, let's say for sake of argument, 6,000 years, then man has been around for 99.999998% of the world's existence. This would make the Lord's statement that man has been on earth since "the beginning of the creation" logical. Conversely, had the earth existed for 4.6 *billion*

AFTER EDEN   by Dan Lietha

AND AFTER MILLIONS OF YEARS OF MUTATIONS, MASS EXTINCTIONS, DEVASTATING DISEASES, AND VIOLENT DEATH, GOD SAW ALL THAT HE HAD EVOLVED AND BEHOLD IT WAS **VERY GOOD!**

119
©AiG 2002

EVOLUTION AND GOD

AnswersInGenesis.org

On this night, little Shelly didn't sleep "very good."

Free Use Policy. © 2015 Answers in Genesis www.AnswersInGenesis.org

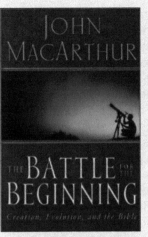

Dr. John MacArthur in his book, "The Battle for the Beginning— Creation, Evolution and the Bible" makes a strong case for a literal interpretation of Genesis chapter one and the six-days of creation. MacArthur contends that destroying a literal interpretation of Genesis one, if consistent, sets one up for an incorrect interpretation of the rest of the Bible.

Denominational scholar Dr. John MacArthur warned against destroying one's interpretation of the Bible by denying the literal truth of Genesis chapters one & two.

years, with man present only the last ~100,000 years, that would have man existing during the last 0.00217% of the world's existence, making Jesus' statement that we have been here since "*the beginning of the creation*" inaccurate and nonsensical. Further, Paul reiterated Christ's statement in Romans 1:20, 21 in declaring that mankind has been able to witness the power of God since "*the creation of the world.*"

Conclusion

Could God have taken 4-5 *billion* years to create the earth if He had chosen? Certainly. In fact, He could have taken 4.6 trillion years or 4.6 nanoseconds. However, the Genesis account of a six literal twenty-four hour day creation is unchangeable. "For in six days the Lord made the heavens and the earth, and on the seventh

day He rested" (Exod. 31:17). One might ask, "Why is a literal interpretation of Genesis important?" Here is why. If one does not correctly understand and interpret the first two chapters of the Bible (*e.g.*, a six literal twenty-four-hour day creation week, a literal first man and first woman, and a literal serpent that tempted the woman), what is to prevent him from being led down the slippery slope of questioning all other literal statements in the Bible? In fact, some Bible scholars, theologians, and Bible students have followed this course and in addition to denying the literal account of Genesis one and two, they now deny the parting of the Red Sea. They deny that Moses actually wrote the Pentateuch. They deny the virgin birth of our Savior. They deny the resurrection of the Christ, and they deny that the New Testament writers wrote by inspiration of the Holy Spirit.

NOTES

_____
_____
_____
_____
_____
_____

Even the denominational author and teacher, Dr. John MacArthur, recognizes and bemoans this fact in the following statement: "In other words, if you reject the creation account in Genesis, you have no basis for believing the Bible at all. If you doubt or explain away the Bible's account of the six days of creation, where do you put the reins on your skepticism? Do you start with Genesis three, which explains the origin of sin...? Or maybe you don't sign on until sometime after chapter six, because the Flood is invariably questioned by scientists, too. Or perhaps you find the Tower of Babel too hard to reconcile with the linguists' theories about how languages originated and evolved. So maybe you start taking the Bible as literal history beginning with the life of Abraham. But when you get to Moses' plagues against Egypt, will you deny those, too? What about the miracles of the New Testament? Is there any reason to regard any of the supernatural elements of biblical history as anything other than poetic symbolism?... If we're worried about appearing 'unscientific' in the eyes of naturalists, we're going to have to reject a lot more than Genesis one through three" (2001, p. 44, emphasis in the original). If understanding the beginnings of mankind isn't crucial to our understanding of the Bible, then nothing else is.

Questions
FILL IN THE BLANKS

1. Other names for theistic evolution include _____
_____.

2. *"Every word of God is _____. Do not _____ _____ His words, lest He rebuke you, and you be found a _____."*

3. The idea that God allowed all non-living matter in the universe to evolve over billions of years may be called _____ _____.

4. Our understanding of our origin should be based on _____ _____ and not on _____ experts.

5. Some theologians view 'nature' as the _____th book of the Bible.

6. *"We have failed to find a single example of the use of the word _____ in the entire Scripture where it means other than a period of ____ _____ when modified by the use of the numerical adjective."*

Yes or No

1. ___ Do the world-class Hebrew scholars at major universities mentioned in this lesson believe that the word "day" in the creation account refers to long ages and not twenty-four-hour solar days?

2. ___ Do the entertainment industry, mass media, and public schools and universities present both sides to the creation-evolution controversy? If not, what side do they lean toward?

3. ___ Does modern evolutionary science teach that the earth is approximately 4.6 million years old?

4. ___ Does the word *yom* in the Bible accompanied by the words "morning and evening" ever mean anything other than a twenty-four hour day?

5. ___ God should have been clearer as to the length of the days in the creation account.

6. ___ Are there other words in the Hebrew that God could have used to indicate ages instead of days?

Short Answer

1. What is the difference between naturalistic and theistic evolution? _____
_____
_____
_____

What are two other names for naturalistic evolution? _____

2. Why do many Bible students compromise their belief of a literal six-day creation week with the day-age theory? _____
_____

3. What did the apostle Paul mean in Romans 3:4 when he said, "Indeed, let God be true but every man a liar"? _____

How does this verse apply to our lesson? _____
_____
_____

4. How is Genesis 1:14 relevant to the day-age controversy? _____
_____

5. How should Christ's statement in Mark 10:16 clarify that the days of creation were twenty-four-hour periods and not ages of billions of years? _____
_____
_____

6. Why is a literal interpretation of Genesis one and two important? _____
_____
_____

Discussion Question in Preparation for Answering Unbelievers and Critics

Your best friend who is a Christian one day sends you an email that says, *"my preacher told me that it's OK to believe that the world is billions of years old as long as you still believe that God created it."* How do you respond? _____
_____
_____
_____

Lesson 5

The Failings of Theistic Evolution (II): Organic (Living) Evolution

Introduction

Lesson four introduced the doctrine of theistic evolution as well as the concept of inorganic evolution, or the evolution of non-living matter in the universe. Although some believers advocate the evolution of non-living matter only, others accept the doctrine of *organic* theistic evolution. This is the idea that all non-living creation as well as all living organic matter arose through evolutionary processes via the consent /direction of God. Dr. R.L. Wysong explained it this way, "... theistic evolution contends that abiogenesis (spontaneous formation of life from chemicals) and evolution (amoeba to man through eons) have occurred, but a creator was instrumental in forming the initial matter and laws, and more or less guided the whole process" (1976, p. 63). (Note: From here forward, the term "theistic evolution" in this lesson will be used exclusively of organic theistic evolution.)

You'll probably never hear a preacher say this is what he's going to do, but...

Free Use Policy. © 2015 Answers in Genesis www.AnswersInGenesis.org

Catholic Pope John Paul II, on October 22, 1996, decreed that macroevolution of man and animals was accepted by the Catholic church (Wikipedia-Public Domain).

The Roman Catholic church has recently embraced theistic evolution as church doctrine. Other proponents of this teaching include the more liberal segments of Judaism, and even a great many in liberal Protestant groups such as Episcopalians, Presbyterians, Lutherans, "Disciples of Christ," and Methodists. Some profes-

Abilene Christian University (supported by some churches of Christ) teaches macroevolution in its science curriculum courses *General Biology II* , and *General Mammalogy. www.acu.edu/catalog/2013_14/courses/ biol.html* One professor at ACU even described the first chapter of Genesis as a *"myth hymn"* (Thompson 2001b).

The humanistic-leaning Time magazine cover for July 23, 2001, "How Apes Became Human."

Bill Nye the so-called science guy (a mechanical engineer) teaches children that, "this thing that happens is called evolution, and it's been going on for billions of years" (Nye, 2011; Photograph, Hs4g, 2010).

Pepperdine University (self-described as a "Christian University" supported by some churches of Christ) teaches macroevolution in its five following science curriculum courses: Principles of Biology, Behavior Mechanisms in Ecology, Biology Senior Seminar, Science as a Way of Knowing, Comparative Animal Behavior. http://seaver.pepperdine.edu/academics/content/2015seavercatalog.pdf

sors in so-called "church of Christ" colleges (*e.g.,* Abilene Christian University, Oklahoma Christian University, and Pepperdine University) have even conceded this doctrine as well (Thompson, 2001b; Jackson, 1984; Gipson, 1999). One professor at Pepperdine, when asked if Adam, Eve, Cain, Abel, Noah, and the worldwide flood were literal events responded by saying, "No they were not real people or real events" (Loveness, 2014). Based on Pepperdine's teaching of evolution in their classes and their refutation of a literal six-day creation week and a literal worldwide flood, one man claims that his brother lost his faith in the word of God (See Loveness, 2014). These individuals are like the dualistic-minded children of Israel scolded by Elijah's searing reprimand, "'How long will you go limping with two different opinions? If the Lord is God, follow him; but if Baal, then follow him.' And the people did not answer him a word" (1 Kings 18:21, RSV).

I. The Views of Others Concerning Theistic Evolution

Many self-proclaimed Christians are pressured into rejecting the clear Genesis account of creation by commingling neo-Darwinian philosophy with the biblical view of man's origin. This should come as no surprise based on evolution being paraded as scientific fact by most secular publications, television and radio programs (*e.g., Time Magazine, PBS, National Public Radio, U.S. News and World Report, Newsweek, Scientific American, Discover, National Geographic, The Learning Channel, The Discovery Channel, Fox News, Reader's*

Digest, Weekly Reader, and *Smithsonian Magazine*). Hollywood even suggests Darwinism in movies such as *Jurassic Park, X-Men, Star Trek, Inherit the Wind,* and the motion picture entitled *Evolution.* Because of societal pressure, individuals often feel coerced into accepting theistic evolution and are, consequently, respected by neither the most devout theologians or the most committed humanists.

To many professed Darwinists, this middle ground is simply ludicrous. Notice the mocking credulity one atheist university professor metes out toward those attempting to marry evolutionary theory with biblical theology: "If a man sets out to build himself a modest home, he does not construct an aquarium, and then remodel it into a chicken coop, and then remodel in into a pigsty, and then remodel it into a dairy, and then into a horse barn... and finally after unnumbered decades into a home for himself. Yet the modernists tell us that God set out to make of the earth a home for a being of spiritual possibilities known as man. So His 'method' of creation was to follow a procedure infinitely more asinine (i.e., absurd, J.G.) than I have just described. For millions of years, such impossible pointless monstrosities as dinosaurs and saber-toothed tigers were the highest type of life on the earth. Life was forced into ten million grotesquely shaped different

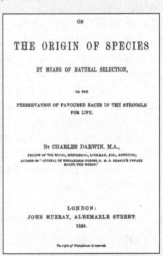

With the publication of Darwin's On the Origin of Species, many religious thinkers began to compromise a literal interpretation of the Bible and the history of creation with Darwinism. (Photocopy public domain via Apologetics Press).

Neo-Darwinism architect, Dr. George Gaylord Simpson, believed that theistic evolution was damaging to science and religion.

molds before it finally was cast in the form of an ape-like creature which passed for man. That method of creation would do discredit to the most brainless idiot who ever lived; it is a madman's method of creation if it is a method at all... If truly there is a God back of evolution and he was intent on the creative evolution of many through bloody struggle, why didn't he get to the point?... If you set out to make a suit of clothes, you don't spend years and years and years evolving masquerade costumes that don't fit you. If an intelligent being set out to evolve a man, He would not spend millions of years evolving grotesque mistakes, animals without a chance to escape extinction" (Dilbeck, 1978, pp. 3, 12).

Further, one of the most esteemed evolutionists of the past century and architect of the modern synthesis of evolution, George Gaylord Simpson of Columbia University, weighed in on his disdain for theistic evolutionists, "The attempt to build an evolutionary theory mingling mysticism and science has only tended to vitiate (i.e., corrupt, J.G.) the science. I strongly suspect that it has been equally damaging on the religious side" (1964, p. 232). Although we would disagree with these gentlemen as to origin of the world, we would wholeheartedly agree that the marriage of biblical theism and Darwinism is unwise.

II. Theistic Evolution's Proposed Mechanisms

We use the word *mechanisms* in the plural because there are many suggested ideas as to how God supposedly directed evolution. Some advocate that God created matter, yet allowed all life forms to naturalistically come into being by unguided evolution. Others claim that God intervened only sparingly throughout this evolutionary period, while still others propose the idea that the evolution of living and non-living matter all occurred by God's directing hand. We will look at three of the most commonly alleged mechanisms of theistic evolution

A. The Day-age Hypothesis. See Lesson four for an overview and examination of this doctrine.

B. The Gap Hypothesis. Although variations of the gap hypothesis exist, the meat of this teaching is that between Genesis 1:1 and 1:2, billions of years transpired, during which evolu-

tion occurred. This idea has been suggested in the popular *Scofield Study Bible* and influenced many believers since 1909. The gap hypothesis, even theologically weaker than the day-age hypothesis, creates a twisted allegorical interpretation of the

The Scofield Study Bible has heavily influenced Protestant belief in theistic evolution since 1909.

Genesis account. The gap teaching claims that in the intervening eons of time between the first two verses of Genesis, God created a fully functional, symbiotic, biologically-active environment with land, oceans, and air teeming with living creatures, which provides an explanation for the fossil record. However, at some point, possibly due to Satan's fall, the earth was corrupted and laid waste and all life destroyed by catastrophe. Then God (Gen. 1:2-27) **re-created** the world along with man. Such teaching, however, makes God's creation of light, evening and morning, the lights in the heavens (*i.e.,* sun, moon, stars, and planets) and the separation of the firmament from the waters, and the emergence of dry land in verses 2-18 preposterous, since the gap theorists claim that all this had occurred millions of years prior to the *re-creation* (McArthur, 2001, p. 75).

C. The Framework Hypothesis. This doctrine says that the days in Genesis one and two have no relationship to any concrete periods of time, but are only a metaphorical, symbolic framework of events, whereby the earth was created. This idea rearranges the events in Genesis chapter one so that days one and four occurred simultaneously (*i.e.,* the creation of light and of the heavenly bodies), days two and five occurred simultaneously (*i.e.,* the separation of the waters from the firmament and the creation of fish and birds), and days three and six occurred simultaneously (*i.e.,* the appearance of dry land and the creation of land-dwelling animals). This teaching strikes at the very core

Many people believe that "young earth creationism" (a literal interpretation of the six-day creation week in Genesis one) is a fairly recent view, popularized by fundamentalist Christians in the 20th century. In his book, *The Great Turning Point*, Dr. Terry Mortenson published part of his doctoral dissertation work that he researched in the U.S. and great Britain. He describes people called the "scriptural geologists" who fought hard for a literal six-day creation account from Genesis based on a catastrophic interpretation of the geologic record. He described how believers and churches compromised to the "uniformitarian" belief that the earth was millions of years old.

In *Creation Compromises*, a Ph.D. scientist outlines numerous compromises that Christians have made over the years regarding the doctrine of origins. Topics covered include alternative hypotheses to a literal six-day creation week, theistic evolution, the "double revelation" theory, the Bible and the age of the earth, the "gap theory," the "day-age," theory, the "framework hypothesis," etc. This 428-page book is available for purchase or FREE in PDF format from www.apologeticspress.org.

of biblical interpretation and relegates the Scripture to a book of indiscernible allegories, opening it up to sundry, diverse, and contradictory interpretations. With respect to this inevitable consequence, Dr. John McArthur states, "What old-earth creationists... are doing with Genesis 1-3 is precisely what religious liberals have always done with all of Scripture—spiritualizing and reinterpreting the text allegorically to make it mean what they want it to mean. It is a dangerous way to handle Scripture. And it involves a perilous and unnecessary capitulation to the religious presuppositions of naturalism—not to mention a serious dishonor to God....Give evolutionary doctrine the throne and make the Bible its servant, and you have laid the foundation for spiritual disaster" (*Ibid.,* pp. 20, 26).

III. Why Christians Should Not Accept Theistic Evolution

In addition to the previous arguments addressing the three proposed mechanisms of theistic evolution, let's now consider further evidence as to why the Bible believer should not succumb to this doctrine.

A. Theistic evolution advocates that Genesis chapters one and two are totally symbolic, or that parts are literal while other parts are allegorical. The Scriptures, however, teach that the Genesis record of creation is a literal and historic account (Exod. 20:11; 31:17; Mark 10:6; Rom. 1:20, 21). "Nothing about the Genesis text itself suggests that the biblical creation account is merely symbolic, poetic, allegorical, or mythical" (*Ibid.,* p. 18). The evolutionist's twisted poetic interpretation of the Genesis account often extends

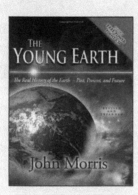

Written by Dr. John Morris (Ph.D) from the Institute for Creation Research. From the book's description, "Scientifically and biblically, the evidence is overwhelming that our planet is not billions of years old. Read why the issue of an old-earth has been devastating for the Church, and why compromise on this issue has rendered much of Bible teaching ineffective. Covers a wide range of topics, from geology to theology. Heavily illustrated. A powerful resource, it also includes a CD-ROM with PowerPoint presentations that illustrate such key concepts as salt levels in the oceans, the age of the atmosphere, the accumulation of ocean sediments, and much more."

on into chapter three, allegorizing particulars such as Satan's temptation of Eve while in the form of a serpent. "Indeed, most theological liberals do insist that the talking serpent in chapter three signals a fable or a metaphor, and therefore they reject the passage as a literal and historical record of how humanity fell into sin... If we cannot believe the opening chapters of Scripture, how can we be certain of anything the Bible says?" (*Ibid.*, pp. 21, 29).

B. The Scriptures teach that creation occurred instantaneously at the command of God (Gen. 1:3, 6, 9, 14, 20, 24; Pss. 33:6-9; 148:1-5; Heb. 11:3). Theistic evolution teaches that the earth and all that it contains were created over *billions* of years of unhurried evolutionary progression.

C. The Bible teaches that the earth was created on day one and the stars were created on day four (Gen. 1:1-5, 14-19). Theistic evolution teaches that the stars were formed billions of years *before* the earth.

D. Although Genesis 1:16-19 teaches that the sun and the moon were both created on day four, theistic evolution teaches that the sun was created millions of years prior to the moon.

The giant panda eats only vegetation although it has carnivore-like claws and teeth (Manyman, 2011).

E. The Bible teaches that animal life and plant life were created on days three and six, respectively (vv. 10-13, 24, 25). Theistic evolution teaches that plant and animal life evolved alongside one another over millions of years.

F. The Bible teaches that both birds and fish were created on day five (vv. 20-23). Theistic evolution, on the other hand, advocates that fish preceded birds by millions of years.

G. Although Genesis 1:20-25 teaches that birds were created first and then reptiles were created the following day, theistic evolution avows that reptiles evolved millions of years prior to birds.

H. The Bible teaches that plants and animals were created fully mature and of reproductive age (vv. 11, 12, 20-25). Theistic evolution says that plants and animals evolved and grew from infantile, embryonic to mature states over millions of years.

I. Genesis chapter one professes that plants and animals reproduce, bearing offspring after their own kind (Gen. 1:11, 12, 21, 24-25). Theistic evolution proclaims that over millions of years, plants and animals produced descendants of *another* kind.

J. The Bible teaches that Adam was the first man (Gen. 1:2; 1 Cor. 15:45). Theistic evolution teaches that the first man existed long before Adam and may have been either *Homo erectus, Homo habilis,* or *Australopithecus afarensis* (Lucy). Notice how Dr. John McArthur views this mangled interpretation: "If everything around these verses is handled allegorically or symbolically, it is unjustifiable to take those verses in a literal and historical sense. Therefore, the old-earth creationists' method of interpreting the Genesis text actually undermines the historicity of Adam. Having already decided to treat the creation account itself as myth or allegory, they have no grounds to insist (suddenly and arbitrarily, it seems) that the creation of Adam is literal history. Their belief in a historical Adam is simply inconsistent

The Jamaican fruit bat eats only vegetation although it has carnivore-like teeth (Photo: Public Domain, CDC)

with their own exegesis of the rest of the text" (*Ibid.*, p. 19).

K. Genesis 2:18-23 declares that woman was fashioned from a rib taken from the side of man. Theistic evolution maintains that women evolved alongside man over millions of years.

L. Theistic evolution says that God's creative work can be seen in the evolution of all life forms and even continues to this day. The Scripture says that God ceased from all his creative work "from the foundation of the world" (Gen. 2:2; Heb. 4:3, 10).

M. All animals were, at first, created to be vegetarians (Gen. 1:29-30). In spite of this, theistic evolution pronounces that the first animals, from the dinosaurs to the great cats, were carnivorous creatures, ravaging and devouring one another. Before one objects to the fact that carnivores were originally herbivores, remember that there are many strict vegetarian animals in the world today with sharp, carnivore-like teeth (*e.g.*, fruit bats and giant panda bears).

Conclusion

The Christian must be careful not to surrender sound biblical faith and a singular understanding of God's word to become a hostage "of science falsely so-called" (1 Tim. 6:20, KJV). Jesus said, "Heaven and earth will pass away, but My words will by no means pass away" (Matt. 24:35). When an individual circumvents God's literal teaching, beginning with the Genesis account, there no longer remains a true and absolute standard whereby he/she will accept the rest of the Scripture as literal. As the Psalmist lamented, "If the foundations are

NOTES

destroyed, what can the righteous do?" (Ps. 11:3). For a fuller treatment of the days of creation, see Appendix C1.

Questions
TRUE OR FALSE

1. ___ Few religious groups teach theistic evolution today.

2. ___ No one who claims to a be member of the church of Christ teaches theistic evolution.

3. ___ Theistic evolutionists garner no support from either the most dedicated humanists or the strictest theologians.

4. ___ The Bible teaches that the first plants and animals were created in an infantile state and later grew to maturity.

5. ___ Elisha rebuked the children of Israel for their vacillating spirit and said, "*How long will you go limping with two different opinions?"*

Multiple Choice
Circle all correct answers

1. The idea that billions of years transpired between Genesis 1:1 and 1:2 is commonly called the (a) Day-age Hypothesis; (b) Gap Hypothesis; (c) Framework Hypothesis; (d) Gen. 1:1, 2 Hypothesis.

2. The idea that each day in the Genesis account of creation represents long eons of time is commonly called the (a) Day-age Hypothesis; (b) Gap Hypothesis; (c) Framework Hypothesis; (d) Long Day Hypothesis.

3. The idea that Genesis 1 presents a symbolic set of events whereby the world was created with no respect to time or order is commonly called the (a) Day-age Hypothesis; (b) Gap Hypothesis; (c) Framework Hypothesis; (d) Allegory Hypothesis

4. What event from Genesis three did we state that some evolutionists interpret symbolically rather than literally? (a) Eating the fruit of the forbidden tree; (b) The cherubim with the flaming sword that guarded the way to the tree of life; (c) God's curse upon man and woman because of their sin; (d) The serpent that tempted Eve; (e) God forming woman from the rib of man.

5. Theistic evolution teaches that the stars were created billions of years ahead of the earth. However, the Bible teaches that the earth was created on day one and the stars were created on day (a) 5; (b) 4; (c) 3; (d) 2.

6. Popular influential media outlets that often teach evolution include (a) *National Geographic;* (b) *PBS;* (c) *Weekly Reader;* (d) *Time Magazine.*

Short Answer

1. Explain the difference between organic and inorganic theistic evolution? _____

_____

_____

Is either idea supported by the Bible? _____

2. Name some passages outside the book of *Genesis* that show that the chronology of the

creation account in Genesis is to be taken literally and not figuratively. _____
_____

3. If one of the events from Genesis listed in multiple choice question #4 above is interpreted figuratively, is there anything that would prevent an individual from interpreting the four other events figuratively? _____

4. If God created the world through billions of years of evolution, which the evolutionist says continues to this day, would God be finished with His creative work?_____
What Scriptures have bearing upon this question? _____

5. What verse in the New Testament tells us to beware of *"science falsely so-called"*? _____
_____

6. Do you believe these last two lessons might be helpful for a theistic evolutionist? _____
_____
Do you know any theistic evolutionists? If so, how might you convey this information to them?
_____
_____

Discussion Question in Preparation for Answering Unbelievers and Critics

You are at a family reunion and a relative, who is studying to be an "ordained minister" in a denomination, tells you that he has learned that all educated religious people believe that evolution occurred by divine providence. He then asks you your view on creation. How do you respond? _____
_____
_____
_____
_____
_____

Lesson 6

Defeated Neo-Darwinian Dogmas (I): Irreducible Complexity, the Intelligent Design Movement, and Radiometric Dating

Introduction

The next three lessons will expose the irrationality of many long-held Neo-Darwinian doctrines. Once considered sacred icons to their movement, the disintegration of these spurious evolutionary proofs has disillusioned many Darwinists. Macroevolution is not firmly planted on a bedrock of irrefutable ideas, as its proponents would have you believe. Unfortunately, many have discovered that supporters of naturalism will go to almost any length, even deception, to defend their cause. When one unshrouds these evolutionary misrepresentations, however, the majority of the Neo-Darwinian community rallies together to discredit the messenger. Dr. Michael Behe admits, *"The Darwinists have a lot of good psychological tricks at their disposal"* (Weiland, 1998, 17). Evolutionist, Dr. Henry Lipson, University of Manchester, Head of the department of physics, and Fellow of the Royal Society said, "evolution became in a sense a scientific religion; almost all scientists have accepted it and many are prepared to 'bend' their observations to fit with it" (Lipson, 1980). Such fraudulence and cover-ups bring to mind the teaching of Jesus concerning man in his day: "And this is the condemnation, that the light has come into the world, and men loved darkness rather than light, because their deeds were evil. For everyone practicing evil hates the light and does not come to the light, lest

The Biochemical Challenge to Evolution

DARWIN'S BLACK BOX

"No one can propose to defend Darwin without meeting the challenges set out in this superbly written and compelling book." —David Berlinski, author of A TOUR OF THE CALCULUS

MICHAEL J. BEHE

his deeds should be exposed" (John 3:19-20). We will discuss two untenable evolutionary arguments in this lesson.

I. Gradual Changes Do Not Accumulate into Complex Biological Systems (Irreducible Complexity and the Intelligent Design Movement)

The first failed philosophical icon of evolution we will examine has probably received more attention in recent years than any other. This teaching goes something like this, "Well organized, complex, and intricately interdependent biological and biochemical systems have evolved over millions of years, adding complexity to the system one piece at a time. Biological systems have, thus, progressed from rather simplistic to perplexingly sophisticated and elaborate living machines. Piece by piece, the organism was edified to create the masterpiece we see today." Such a statement assumes that biological systems are "reducibly complex." That is to say that an entire biological system could continue functioning even if the system were disassembled one piece at a time. Even Darwin admitted that for

Professor and biochemist, Dr. Michael J. Behe, authored Darwin's Black Box, which poked holes into Darwinian evolution through an explanation of the irreducible complexity of biological systems. (Copyrighted photo. Used by permission.)

evolution to have occurred, biological systems must be **reducibly** and not **irreducibly complex**. He stated, "If it could be demonstrated that any complex organ existed which could not possibly have been formed by numerous, successive, slight modifications, my theory would absolutely break down" (1872, 146).

Professor Michael Denton, Senior Research Fellow in the Biochemistry Department at the University of Otago, Dunedin, New Zealand. Authored the book, Evolution: A Theory in Crisis, 1985, which started the modern Intelligent Design Movement. His book convinced Professor Michael Behe and many other evolutionists that Neo-Darwinism was wrong (Photo: Courtesy of successfulstudent.org)

U.C. Berkely Professor, Dr. Philip Johnson, who organized the modern Intelligent Design Movement in the United States, and penned the book, Darwin on Trial, 1991 (Photo: Courtesy of Greg Schneider).

The book that started the modern Intelligent Design Movement, authored by New Zealand biochemist, professor Michael Denton.

The modern Intelligent Design Movement was started in 1985 by biochemist Dr. Michael Denton and his book, *Evolution: A Theory in Crisis.* This movement was organized in the United States by Dr. Philip Johnson who authored, *Darwin on Trial* in 1991. Following this publication, Lehigh University professor of Biochemistry, Dr. Michael J. Behe, in 1996, published his groundbreaking work, *Darwin's Black Box,* which leveled a striking blow to the very heart of evolution and reducible complexity. Behe countered the idea of reducibly complex biological mechanisms with his explanation of *"irreducibly complex systems."* That is to say, living biological and biochemical systems are so inextricably reliant upon all their interdependent components that removing even one essential piece creates a total melt down of the mechanism and renders the system useless. Thus, the complex system is **irreducible** and not **reducible**. Behe defined the term this way: "By irreducibly complex I mean a single system composed of several well-matched, interacting parts that contribute to the basic function, wherein the removal of any one of the parts causes the system to effectively cease functioning. An irreducibly complex system cannot be produced directly (that is, by continuously improving the initial function, which continues to work by the same mechanism) by slight, successive modifications of a precursor system, because any precursor to an irreducibly complex system that is missing a part is by definition non-functional" (1996, 39). How

The Discovery Institute is the hub for the Intelligent Design Movement in the U.S. (www.discovery.org)

profound yet how simple. For a comparison/contrast and integration of the Intelligent Design Movement and Biblical Creationism (which affirms a literal twenty-four-hour, six-day creation week in Gen. one) see **Appendix C2.**

To learn that living organisms are constructed this way creates no difficulty for the creationist, because an omnipotent Creator is not restricted to create biological systems incrementally over time. An all-powerful creative force can create such a mechanism instantaneously and fully functional with all the required parts if He so desires. The thought of this occurring is at once awe-inspiring and praise-worthy as was eloquently stated by David and his son Solomon. "I will praise You, for I am fearfully and wonderfully made; Marvelous are Your works, And that my soul knows very well" (Ps. 139:14). "As you do not know what is the way of the wind, Or how the bones grow in the womb of her who is with child, So you do not know the works of God who makes everything"

Ben Stein's movie, Expelled: No Intelligence Allowed, reveals how scientists in the U.S. are being persecuted for believing Intelligent Design.

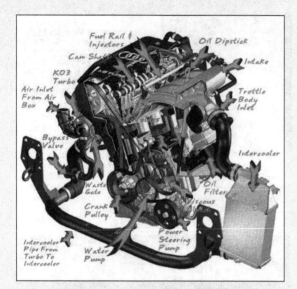

The automobile engine. An example of an irreducibly complex design in non-biological systems. (Public domain via Greg Farber.)

(Eccl. 11:5). Let's look at two man-made machines that are much simpler than those of the biological sort. This should help us get a better feel for the concept and dynamism of irreducible complexity.

A. The automobile engine. Let's take the modern internal combustion engine and inquire as to which part could be removed while allowing the motor to continue functioning. What would happen if you removed the engine's computer system, or the carburetor or throttle body, or the radiator, or the cam shaft? Removing any one of these parts would, of course, result in a non-functional engine. But what if you removed one of the smaller parts? The spark plugs, or the valves, or the rocker arms, or the push rods? The same thing would occur. Now, what if you removed one of the smallest and seemingly insignificant parts such as a head gasket, the rear main seal, the oil pan plug, or even a cooling hose? Remove one part and the engine is effectively dead; thus, an automobile engine is an irreducibly complex machine. Yet, Darwinists expect us to believe that biochemical systems functioned over billions of years with critical parts missing. Now let's look at an even simpler mechanism.

B. The mousetrap. In his book, Dr. Behe uses the example of a mousetrap (1996, 42-47). It is a simple contraption, yet composed of irreducibly complex parts. The common mousetrap is made up of (A) a small wooden platform, (B) the metal bar or hammer that snaps the mouse, (C) the coiled wire spring that, when bent, charges the hammer, (D) the metal catch that trips when a mouse presses it, and (E) a metal bar that holds the hammer back and attaches to the metal catch. A mechanism this simple certainly should be able to function if one part was removed—shouldn't it? But which part can you remove? **None.** They are all essential. This is another example of a machine that is irreducibly complex. The most complex machines known to man, however, are not the ones created by him. They are those present in the natural world—plants, animals, and microorganisms. Let's look at some of them.

C. The human eye. The most delicately designed inventions of man will never equal the complexity or beauty of those created by Jehovah. God through Jeremiah said, "I have made the earth, the man and the beast that are on the ground, by My great power and by My outstretched arm, and have given it to whom it seemed proper to Me" (27:5). Vision in the human eye is only possible through a series of complex and irreducibly complex biochemical reactions. Biochemistry is the study of chemical reactions in living organisms. Although many of the following terms might be foreign to you, let us look at the most basic of biochemical processes that are required to produce one glimmer of sight. (1) Light strikes the retina of the eye→ (2) Light photons are absorbed by the molecule 11-*cis*-retinal→ (3) This molecule rearranges itself into *trans*-retinal (this molecular rearrangement occurs in picoseconds [a picosecond is 10^{-12} of a second or the time it takes light to travel the width of a

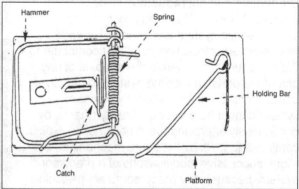

The mousetrap. The simplest of machines is still irreducibly complex (From Behe, 1996. Used with permission).

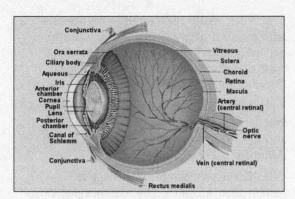

The human eye. A biological machine of irreducibly complex design. (Permission for use granted by The Discovery Fund for Eye Research, Los Angeles, CA.).

human hair])→ (4) This forces rearrangement in the protein rhodopsin to metarhodopsin II→ (5) Rhodopsin can now interact with GDP-associated transducin→ (6) Transducin then binds GTP→ (7) GTP-transducinrhodopsin then binds to the protein phosphodiesterase→ (8) Phosphodiesterase cleaves and lowers the concentration of cGMP→ (9) The sodium ion channel is then closed, concentrating sodium cations→ (10) This creates an imbalance across the cell membrane initiating a signal through the optic nerve to the brain, creating vision (Behe, 1996, 18-21). I imagine you never knew so much occurred every time you opened your eyes. This is only a hint of what actually occurs in vision and it is indisputable that this system is irreducibly complex. This process not only requires all of these integrated parts, but it also demands that they perform in fine precision in an amazingly short amount of time. And what happens if just one of the ten steps listed above fails to work? The results are blinding.

How, then, do the Darwinists explain the evolution of the eye? They can't. The key point?? Neo-Darwinism has no mechanism of how biochemical systems evolved, since they are irreducibly complex. An evolution-supporting Professor Emeritus of Biochemistry at Colorado State admitted this, saying, "There are presently no detailed Darwinian accounts of the evolution of any biochemical or cellular system, only a variety of wishful speculations" (Harold, 2001). At least there is no consensus as to how such a miraculous event could have taken place. It is inconceivable that the evolution of this biological system occurred over millions of years until finally, the last piece was added to

the puzzle and, **POOF,** sight appeared. Non-functional and useless eyes existing on animals for millions of years? Who can believe it? Remember Darwin's statement? "If it could be demonstrated that any complex organ existed which could not possibly have been formed by numerous, successive, slight modifications, my theory would absolutely break down" (*op. cit.*). And although Charles Darwin had no inkling as to these elegant biochemical processes in 1859, notice his eerily ominous statement, "To suppose that the eye with all its inimitable contrivances for adjusting the focus to different distances, for admitting different amounts of light, and for the correction of spherical and chromatic aberration, could have been formed by natural selection, seems, I freely confess, absurd in the highest degree" (1872, 143). It is much more logical to suppose that an intelligent being created the eye. How this should move us to reverence and praise. "Many, O Lord my God, are Your wonderful works which You have done; and Your thoughts toward us cannot be recounted to You in order; if I would declare and speak of them, they are more than can be numbered" (Ps. 40:5).

D. The bacterial flagellum: majestic miniature molecular motors. This final example couples biochemistry with the more mechanistic flagellum, present on some bacteria. A flagellum is a long whip-like appendage protruding from the bacterial cell membrane. It is used somewhat like a boat propeller to provide thrust for a bacterium, producing mobility in fluid. What is fascinating is that the flagellum can rotate over 1,000 times per second by means of a tiny electrical motor

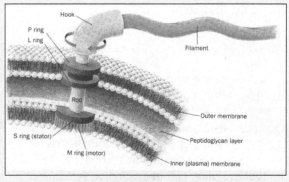

The bacterial flagellum. A microbiological machine of complex design (Illustration by Irving Geis. Courtesy of the Howard Hughes Medical Institute. Rights owned and administered by the Howard Hughes Medical Institute. Reproduction by permission only).

composed of various proteins. Long before mankind invented the motor or the wheel, these parts already existed in bacteria. Some of the flagellar motor parts that correspond to man-made electrical motors (man-made parts in parenthesis) include a rod (drive shaft), a hook (universal joint), L and P rings (bushings/bearings), S and M rings (rotor), and a C ring and stud (stator). These motors require ~40 proteins to function. It was also discovered that, unlike muscles that require molecules of energy to perform work, the flagellum derives energy by acid flowing through the bacterial membrane (Behe, 1996, 70-72). Other features of the bacterial motor include self-assembly and repair, the ability to rotate in forward or reverse motion, and the capability of reversing this direction within ¼ of a turn, while rotating at up to 100,000 RPM (Safarti, 2002, 168). In fact, the 1997 Nobel Prize was awarded to Boyer and Walker for discovering the energy-creating mechanism of these microscopic motors. If one of the dozens of complex parts were to be removed from this motor, it would absolutely cease to function. Again, it is beyond the scope of logic to conclude that the gradual evolution of the flagellum occurred over eons of time until millions of years later the last bearing was added, the flagellar motor was for the first time functional, and the bacterium whizzed off with its newly-evolved motility. Who can believe it? In addition to this, there are thousands of molecular machines in a single human body. Where did they come from?

Evolutionist, U.C. Berkeley paleontologist, Dr. Kevin Padian, who, although believing macroevolution to be true, casts doubt on the mechanism of Neo-Darwinism (Photo: wiscatheists.blogspot.com).

Many other irreducibly complex systems in biology and biochemistry could be discussed such as (1) the clotting of blood, (2) lobster eyes with unique square reflecting geometry, (3) cellular transport, the (4) mechanism of antibodies and antigens, or something as simple as (5) the human knee joint that requires over fifteen irreducibly complex parts to function. Evolution cannot account for the gradual evolution of these complex systems in nature in light of their irreducible complexity.

D. Four world renowned evolutionists who admit that gradual genetic mutations do not gradually accumulate to cause macroevolutionary changes (i.e., Neo-Darwinism is wrong).

1. "I have seen no evidence whatsoever that these [evolutionary] changes can occur through the accumulation of gradual mutations" (Margulis, 1991).

2. "Darwin's claim of 'descent with modification' as caused by natural selection is a linguistic fallacy... "The source of purposeful inherited novelty in evolution, the underlying reason the new species appear, is not random mutation..." (Margulis, 2010).

3. "New mutations don't create new species; they create offspring that are impaired" (Margulis, 2006).

4. "The experimental results have been available for the last thirty-five years, but have been ignored or silenced to avoid creating cracks in an edifice based on randomness and selection" (Lima-de-Faria, 2010).

5. "The edifice of the Modern Synthesis [i.e., Neo-Darwinism, JBG] has crumbled, apparently, beyond repair... all major tenets of the Modern Synthesis are, if not outright overturned, replaced by a new and incomparably more complex vision of the key aspects of evolution. So, not to mince words, the Modern Synthesis is gone. What's next?" (Koonin, 2009).

6. "How do major evolutionary changes get started? Does anyone still believe that populations sit around for tens of thousands of years, waiting for favorable mutations to occur (and just how does that happen, by the way?), then anxiously guard them until enough accumulate for selection to push the population toward new and useful change? There you have the mathematical arguments of Neo-Darwinism that Waddington and others rightly characterized as 'vacuous' [i.e., empty, useless, lacking intelligence, JBG]" (Padian, 1989).

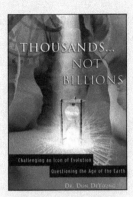

This book (and DVD) summarize eight years of research by the Institute for Creation Research (ICR) and a team of scientists, who used radiometric dating to show the earth is thousands, not billions of years old.

II. Radiometric Dating is Based on Unprovable Assumptions.

The second debunked Darwinian dogma we will examine is the belief that radiometric dating can prove that the earth is billions of years old, during which time evolution occurred. Radiometric dating utilizes readings from unstable radioactive isotopes in rocks and other material in order to extrapolate back to the supposed origin of the substance and provide an alleged age. There are many problems associated with these processes. First of all, carbon-14 (^{14}C) dating (one of the most cited methods of radiometric dating) can only attempt to provide estimated ages of organic materials in the realm of **thousands** and not **billions** of years. Evolutionist, Sheridan Bowman, says 40,000 years is roughly the maximum age that can be measured by the ^{14}C method (1990, 37). Its validity, then, is not germane to a discussion with evolutionists who claim that millions of years were necessary for evolution to occur. Interestingly, although coal was supposedly formed hundreds of millions of years ago if the earth has really been around that long. Coal is commonly found containing ^{14}C, which should have dissipated millions of years ago (Baumgardner *et al.*, 2003; Lowe, 1989). This shows the coal is thousands not billions of years old. This demonstrates that these fossil fuels are in the neighborhood of thousands and not millions of years old.

Additionally, diamonds, the hardest substance known to man, which cannot be internally contaminated, are supposed to be billions of years old containing no more Carbon-14. Nevertheless, ^{14}C has been found inside diamonds by secular scientists, showing that they are thousands and not billions of years old (Taylor and Southon, 2007). [**Note:** See Appendix F for 90 ancient geological samples that are supposedly

millions of years old and thus should contain no more C-14. However, they are now shown to contain C-14 and be less than 40,000 years old (Bowman, 1990) by C-14 dating]. Further, Dr. Sheridon Bowman's 1990 book states, "Radiocarbon is not quite as straightforward as it may seem. The technique does not in fact provide true ages, and radiocarbon results must be adjusted (calibrated) to bring them into line with calendar ages" (back cover).

Methods employed to provide dates of supposed millions of years are based on unprovable assumptions. For example, the isotope uranium-238 (^{238}U), used in radiometric dating, often provides ages of supposed millions of years. The procedure works by determining the amount of radiometric decay the parent element (^{238}U) undergoes on its way to becoming the daughter element, lead-206 (^{206}Pb). The problem is that to standardize the method, three unknowns must be **assumed.** First, it must be assumed that when the ^{238}U was first formed in the rock, there was absolutely no ^{206}Pb present. How can anyone know this? They can't. Second, one must assume that neither the parent element (^{238}U) nor the daughter element (^{206}Pb) has decreased in mass over time. Quite to the contrary, some evidence suggests that, under certain conditions, these radioactive elements will migrate into surrounding rock and/or solubilize in water. Third, it must be assumed that the decay rates of these radioactive elements have remained constant and have not fluctuated over time, although some evidence suggests otherwise (Jackson, 1983, 3-5). Thus, radiometric dating can only truly determine the

Mt. St. Helens new lava dome, which returned radiometric dating ages of between 500,000 and 3 million years old (Photo: USGS).

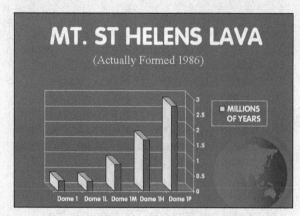

Five different dates assigned through radiometric dating to 11-year-old rock samples taken from five locations on the lava dome of Mt. St. Helens. (Graph courtesy of Don Patton and Steve Rudd. bible.ca).

concentration of isotopes within rock and not its actual age. Examples of radiometric dating providing patently false readings are replete. One example is wood found in Tertiary basalt in Australia dated at 45,000 years by the ^{14}C method and later dated at 45 million years old by the radiometric potassium-argon technique (Snelling, 1998, 24-27).

Another example of the dubious nature of radiometric dating was seen when rock specimens were sampled from five locations on the new lava dome on Mount St. Helens in Washington state. Although the rock was created only eleven years earlier, radiometric dating by the potassium-argon method assigned five dates between 0.5 and almost 3 million years. One paleochronology group tested 8 dinosaur bones found in

Texas, Alaska, Colorado, and Montana by carbon-14 dating and found the ages were between 22,000 and 39,000 years old (Fischer, 2012).

Conclusion

The above-mentioned neo-Darwinian doctrines only raise more questions as to the mechanism, or the possibility, of evolution. Although some of the world's brightest minds have devised these evolutionary ideas, they have clearly revealed themselves to be individuals who would rather use their talents to support the dubious doctrine of Darwinism than to uncover the truth. They are, as those described by the apostle Paul, "Always learning but never able to acknowledge the truth" (2 Tim. 3:7).

An illium bone of an Acrocanthosaurus that was radio carbon dated at only 19,000 years old. This was from an animal that, according to evolution, became extinct over 100 million years ago. (Photo courtesy of Don Patton and Steve Rudd. www.bible.ca).

NOTES

Questions
YES OR NO

1. ___   Are all Darwinian assertions supported by solid facts?

2. ___   Evolutionists teach that complex biological organisms evolved slowly over millions of years until they were complete.

3. ___   Does irreducible complexity show that organisms could not have evolved gradually until their biochemical systems were fully and functionally complete?

4. ___   Did Solomon say, "I will praise You, for I am fearfully and wonderfully made"?

5. ___   Carbon-14 dating can provide the ages of organic materials up to one million years.

6. ___   Peter said that some are, "always learning but never able to acknowledge the truth."

Fill in the Blank

1. Dr. Michael Behe said, "The Darwinists have a lot of good_____ _____ at their disposal."

2. Charles Darwin said, "If it could be demonstrated that any _____ _____ existed which could not possibly have been formed by numerous, successive, _____ _____, my theory would absolutely break down"

3. We used two examples of man-made irreducibly complex systems: the _____ and the _____.

4. We provided six biological illustrations of irreducible complexity, including the _____ and the _____.

5. David said, "Many, O Lord my God, are Your _____ _____ which You have done; and Your thoughts toward us cannot be _____ to You in order; if I would declare and speak of them, they are more than can be _____"

6. Three unprovable assumptions that must be made when using radiometric dating are: _____
_____
_____
_____
_____

Short Answer

1. What are two reasons Jesus gives in John three for people not wanting to come to the light? _
_____
_____

2. Explain the difference between reducible and irreducible complexity. _____
_____
_____

3. What lesson could evolutionists learn from Solomon's teaching in Ecclesiastes 11:5? _____
_____
_____

4. What other examples of irreducible complexity among human inventions can you think of? ___
_____
_____

5. What other examples of irreducible complexity in nature can you think of? _____
_____
_____

6. Explain how radiometric dating proves that coal and diamonds are only thousands and not
millions or billions of years old. _____
_____
_____

Discussion Question in Preparation for Answering Unbelievers and Critics

It is your freshman year of college and you are working on a group project in biology. One of your colleagues remarks, "Don't you think it's amazing that such complex biological features evolved in these animals over millions of years?" Your response is what?_____
_____
_____
_____
_____
_____
_____

Defeated Neo-Darwinian Dogmas (II): Darwin's "Tree of Life," Animal Homology, Homology, Embryology, and the Peppered Moth

Introduction

Far too often in history, Darwinian proponents have resorted to trickery or falsehoods, in order to support their false premise, as we will see in the next several lessons. The Bible warns of those who use these tactics to pervert the truth and deceive the unsuspecting masses (Acts 13:9-11; 2 Cor. 4:2; 11:13, 14; Eph. 4:14; Col. 2:8). In the face of contradicting facts, the following three ideas continue to be used to support macroevolution in public education textbooks.

I. Darwin's "Tree of Life" Has Been Cut Down

Darwin's "Tree of Life" was the only illustration in his 1859 book. The "Tree of Life" was a metaphor meant to replace the Biblical tree of life as a symbol of man's origin. Darwin argued that the evolution of all animals started as a tree from a trunk. Then an unbroken succession of limbs in the tree represented the evolution of each species of animal into the next (See the first figure). Does the scientific evidence point to such an evolutionary branching pattern?

For years it has been known that Darwin's "Tree of Life" presented in his book, *Origin of Species* is incorrect. Most recently, this conclusion has been based on DNA evidence (Lake,

1999; Syvanen and Ducore, 2010). Nevertheless, use of the "tree of life" model is still presented in high school and college text books. Evolutionist Brian Goodwin of Open University, U.K. said, "So Darwin's assumption that the tree of life is a consequence of the gradual accumulation of small hereditary differences appears to be without

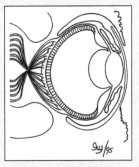

The octopus eye, bearing remarkable similarity to the human eye, cannot be explained by evolution. (Used by permission from the BIODIDAC project).

significant support. Some other process is responsible for the emergent properties of life, those distinctive features that separate one group of organisms from another" (Goodwin, 1995a, p. x). In the *New Scientist,* it was admitted, "For a long time the holy grail was to build a tree of life. But today the project lies in tatters, torn to pieces by an onslaught of negative evidence. Many biologists now argue that the tree concept is obsolete and needs to be discarded. We have no evidence at all that the tree of life is a reality... the evolution of animals and plants isn't exactly like a tree" (Lawton, 2009). One study looked at 2,000 genes from six different animals. The *New Scientist* reported, "In theory, he should have been able to use the gene

As described in this special issue, "Darwin's Tree of Life," "lies in tatters, torn to pieces by an onslaught of negative evidence" (Lawton, 2009).

The sketch of Darwin's tree of life (illustration taken directly from his first book) has been debunked by genetic evidence that shows that lines of supposed evolutionary ancestors have a very different genetic makeup, as determined by DNA (Lake, 1999).

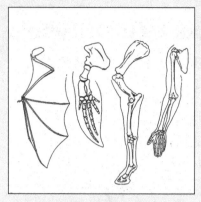

Similarities among vertebral limbs supports the idea of a common creator rather than than a common ancestor. Left to right: Bat, Whale, Horse, Human. (Copyright Jody F. Sjogren 2000. Used with permission.)

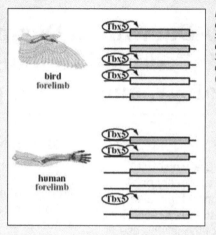

In this illustration, the Tbx5 gene simply acts as a "selector" gene or a "genetic switch," activating the actual downstream developmental genes (pointed to with arrows). Scientists still don't understand what these developmental genes are or how they function (Photo credit: Carroll et al., 2004).

sequences to construct an evolutionary tree showing the relationships between the six animals. He failed. The problem was that different genes told contradictory evolutionary stories" (Lawton, 2009). That's pretty strong language. Did the authors of the study agree? They said, "We've just annihilated the tree of life. It's not a tree anymore, it's a different topology [pattern of history] entirely. What would Darwin have made of that?" (Quoted in Lawton, 2009). Yet, thousands of textbooks around the country continue to teach the Darwinian "Tree of Life" myth.

II. Similarity among Animals (Comparative Anatomy, or Homology)

It is affirmed that because there are similarities among many classes of animals, the evolutionary hypothesis must be true. Regarding animal species and similarity, Charles Darwin in *The Origin of Species* included the following quote, "Is it not powerfully suggestive of true relationship, of inheritance from a common ancestor?" (1872, p. 382). Since that time, similarity among animals (or homology) has been one of the cornerstones of evolutionary teaching. It says that if two very different animals have a very similar feature, they must have both evolved from the same animal. This rule is not strictly followed, however, and it is often broken when two animals, who evolutionists deny evolved from the same ancestor, coincidentally happen to have a similar feature. For example, evolutionists admit that the eye of the octopus

and the eye of the human are remarkably similar. However, the same individuals deny that the eyes in both creatures evolved from a common ancestor. Instead, they argue that it is sheer coincidence that the eyes of the human and of the octopus just happened to evolve into the same type of optical sensory instrument. How, then, do evolutionists determine which animal similarities are indicators of evolution (evolutionarily significant) and which similarities are insignificant? This is done with more faulty and circular reasoning. First the Darwinist says, "We know that a feature in two animals is evolutionarily significant if and only if the two animals had a common ancestor from which this feature evolved." Next, the student asks, "So, how do we know that two animals had a common ancestor?" How does the evolutionist respond? You guessed it. He states something along these lines, "We know that two animals have a common ancestor if they have a similar evolutionarily significant feature." Thus, the octopus eye and the human eye, we are told, are not similar evolutionarily significant features because they have no common ancestor, and they have no common ancestor because their eyes "are not similar evolutionarily significant features." Does that make any sense to you? If not, then you get the point. Scientist Alan Boyden mockingly pointed this inconsistency

Dr. Gavin De Beer questioned why homologous organs were controlled by different genes in 1938 and 1971, and scientists are asking the same questions today. (Dr. De Beer was an embryologist, Director of the British Museum of Natural History, President of Linnaean Society and Winner of the Royal Society's "Darwin Medal").

out many years earlier when he stated, "As though we could know the ancestry without the similarities to guide us!" (1947).

The fact that similarities exist among animals is not in dispute; *why* similarities between animals exist *is*. For the evolutionist, similarities indicate evolution. For the creationist, however, similarities are clear insignias of a common designer. The human eye and the octopus eye are a case in point. The fact that these two very different creatures have the same type of optical structure while many other animals (supposedly more closely related to them by evolution) do not have the same optical instrument cries out for a common *Creator*, not a common ancestor. When the evolutionist points to the similarity in vertebrate limbs and proclaims "evidence of a common ancestor," you should say, "No, this is evidence of a common designer." Consider this example. A contractor builds a house as well as an apartment building out of the same material such as steel girders, bricks, mortar, drywall, *etc.* Although the house and the apartment building are two very different structures, they would have many similarities because they were constructed by the same builder using identical materials. The same is true with animal life. The apostle Paul pointed out that denying this fact is inexcusable, "For since the creation of the world His invisible attributes are clearly seen, being understood by the things that are made, even His eternal power and Godhead, so that they are without excuse" (Rom. 1:20).

The Evolutionists Genetic Embarrassment ("Homologous" Organs Are Coded for by Different Developmental Genes)

Let's grant the evolutionist that organs are homologous, indicating a common ancestor. QUESTION: What must we find in homologous organs that indicate they came from the same common ancestor? ANSWER: The same *genes*, or same genetic code in the DNA. This is a tremendous embarrassment for evolutionists. Many so-called homologous organs have *different* genes controlling their formation. Similar structures with *different* genes says that they must have been created

by a common designer who built the same organs, but used different genetic information to do so. World famous evolutionist Gavin De Beer addressed this and said, "What mechanism can it be that results in the production of homologous organs, the same 'patterns', in spite of their not being controlled by the same genes? I asked this question in 1938, and it has not been answered" (1971). Evolutionists still have no explanation for this genetic coding problem today. Another biochemist professor and physician said, "The evolutionary basis for homology is perhaps even more severely damaged by the discovery that apparently homologous structures are specified by quite different genes in different species. . . There is not a trace at a molecular level of the traditional evolutionary series: cyclosomes to fish to amphibian to reptile to mammal. Incredibly, man is as close to lamprey as are fish! . . .At a molecular level there is no trace of the evolutionary transition from fish to amphibian to reptile to mammal" (Denton, 1986).

The evolutionists claim that because there are "Hox" genes, common to all vertebrate forelimbs (e.g., the Tbx5 gene), this proves common ancestry. NOTE: The Tbx5 gene is a "selector" gene or a "genetic switch." It is not the developmental gene(s) that form each vertebrate forelimb differently. Tbx5 only controls or regulates the actual developmental genes that form the forelimbs (see the second figure on p. 72). Evolutionist Sean Carroll *et al.* (2004) said, "The limb selector gene is expressed in all vertebrate forelimbs, yet these genes regulate different sets of target genes in different lineages. . . these genes likely regulate different sets of target genes to pattern morphological differences between serially homologous forelimbs." Even the National Center for Science Education (which opposes creation being taught in schools) stated, "Development of the vertebrate forelimb is much the same, with the same selector gene Tbx5 controlling different target genes to produce different morphologies in birds and humans, which are homologous in their basic structure, but show variation in the developmental pathways."

Nevertheless, although the "homologous" forelimbs are formed by different developmental genes, and couldn't have a common ancestor, a 2011 review of twenty-two recent biology textbooks found that all twenty-two included the disproven forelimb homology example (Luskin, 2011).

Molecular phylogeny is the use of the genetic codes in animals to prove relationship and common ancestry of a line of animals. Darwin constructed a phylogenetic "tree of life" by which all animals were supposedly related. In recent years, however, genetic testing of animals that should be related to one another by DNA evidence, has proven just the opposite! Animals that are supposed to be related by evolution have very different genes, much to the surprise and embarrassment of the evolutionists. Thus, Darwin's evolutionary "tree of life" has been falsified and evolutionists must use a different model to explain ancestry. Lake *et al.* (1999) said that the excitement that evolutionary ancestry could be proven genetically, "began to crumble a decade ago when scientists started analyzing a variety of genes from different organisms and found that their relationship to each other contradicted the evolutionary tree of life."

As we have shown, Darwinists are quick to demonstrate the similarities in supposedly evolutionarily-related animals. However, they overlook the fact that the same animals have some features that are even more similar to animals they are supposedly unrelated to by evolution. The eyes of the human and the octopus are one example. More examples are as follows:

A. Based on an examination of the amino-acid profile of alpha hemoglobin in blood, the crocodile and chicken are 17.5% similar, the viper and the chicken are 10.5% similar, but the viper and the crocodile (both reptiles) are only 5.6% similar (Wysong, 1976, pp. 394, 395).

B. If the amino-acid sequences of cytochrome C (used by oxygen-requiring animals) are analyzed, one will find that the gray whale is more similar to the duck than another mammal, the monkey. Also, the tuna (a fish) is more similar to the rabbit (a mammal) than another fish, the dogfish; and the turtle is more similar to birds than a fellow reptile, the snake (*Ibid.*; Frair and Davis, 1983, pp. 45-53).

C. The bird wing and the bat wing, although "coincidentally similar," supposedly evolved from different ancestors and are said to be one of the best examples are "convergent" (i.e., "coincidental") evolution.

It should not be difficult to understand why similarities exist among animals that are all composed of the same elements found in the dust of the ground from which God made man. "And the Lord God formed man of the dust of the ground, and breathed into his nostrils the breath of life; and man became a living being" (Gen. 2:7). "In the sweat of your face you shall eat bread till you return to the ground, for out of it you were taken; for dust you are, and to dust you shall return" (Gen. 3:19).

III. Embryology (Haeckel's Embryos)

Another hollow assertion that grew out of the previous argument is that the study of embryology is clear proof of evolution. Embryology is the study of unborn animals in their embryonic state. Charles Darwin, and many evolutionists since then, hypothesized that embryos recapitulate or represent former evolutionary stages of their ancestors, as they grow within the womb. They assumed that embryos of all animals will resemble their evolutionary ancestors

Ernst Haeckel, whose deceptive embryonic drawings continue to promote Darwinian evolution.

at some stage. For example, they might say that at some stage of development all embryos resemble a fish, a salamander, or a lizard. Darwin believed so strongly in this idea that in regard to evolution he said, "Embryology is to me by far the strongest single class of facts in favor of change of forms" (1860). To assist Darwin in his idea, the German biologist Ernst Haeckel, also known as the "apostle of Darwinism in Germany," drew different types of embryos at various stages of development to demonstrate their similarities. It was quickly shown, however, that Haeckel had faked his drawings to make the embryos look more similar to one another than they actually were. In fact, at times he used the same woodcut to print pictures of embryos that

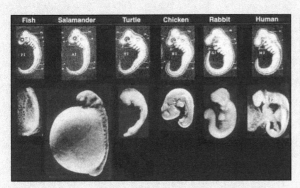

Haeckel's fraudulent embryos (top) and photographs of the actual embryos (bottom).

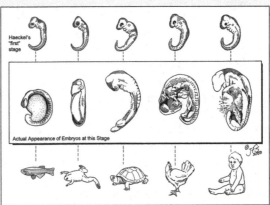

Comparison of Haeckel's fraudulent illustrations with an accurate representation of the appearance of embryos at various stages of development (Copyright Jody F. Sjogren 2000. Used with permission).

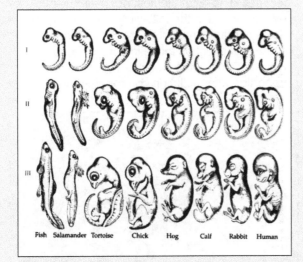

Ernst Haeckel's 1874 fraudulent illustrations of embryonic development. Left to right: Fish, Salamander, Tortoise, Chick, Hog, Calf, Rabbit, Man.

were allegedly from different animals! Even the famed evolutionist, Stephen J. Gould, in a year 2000 world-class article, stated that "Haeckel had exaggerated the similarities" and "in a procedure that can only be called fraudulent—simply copied the same figure over and over again" and "we do, I think, have the right to be both astonished and ashamed by the century of mindless recycling that has led to the persistence of these drawings in a large number, if not a majority, of modern textbooks" (pp. 42-48). In 1997, embryologist Dr. Michael Richardson and a worldwide team of science experts examined in detail Haeckel's drawings and found multiple accounts of fraudulent artistry. The pro-evolution scholarly journal *Science* interviewed Richardson in 2000. In the interview he said, "It looks like it's turning out to be one of the most famous fakes in biology" (Pennisi, 1997, p. 1435). However, over a century after Haeckel's embryos were first proven to be falsified, these drawings continue to be used to teach the hypothesis of evolution in high school and college text books.

Other failed arguments regarding embryos are that, "during a certain stage of development the human embryo has both a tail and gills." The so-called tail-bud was proven to be nothing more than the base of the developing backbone seen in all vertebrates. Evolutionists also espoused that grooves in the neck of the human embryo were really "gill slits." This was later proven false. The embryonic neck ridges are not even slits but are pharyngeal folds in the neck surrounding the supporting arches. The truth of this issue is that actual embryological development does not mimic successive stages of evolution. For instance, evolution teaches that the heart evolved from the blood vessels, and that the teeth evolved after the tongue. However, in human embryology, the heart develops prior to the blood vessels and the teeth prior to the tongue. Examples of some recent college texts that have used the deceptive argument of embryonic recapitulation include the 2002 version of Raven and Johnson's *Biology*, and the 1998 version of Futuyma's well-read advanced college textbook, *Evolutionary Biology*.

Embryologist, professor Dr. Michael Richardson (Leiden University, the Netherlands) and his team reported in the journal Science Haeckel's drawings were fakes (Photo credit: Leiden University, www.leiden.edu)

IV. The English Peppered Moth

Another example used to support evolution in public schools is the peppered moth. The teaching goes like this. Prior to the industrial revolution, the peppered moth in England was typically speckled and light in color. However, a small proportion of the population was darker in color. Due to the vast quantities of industrial pollution and soot that was produced, trees that were light in color suddenly darkened when light colored lichen on the tree bark was destroyed by the pollution. Allegedly, light colored moths could no longer camouflage themselves on tree trunks and were easily picked off of the trees by hungry birds. Further, the darker colored moths were now more easily camouflaged and so the proportion of moths shifted from mostly light colored and very few dark, to mostly dark and very few light colored moths. Supposedly, this was neo-Darwinian evolution occurring right before their very eyes. As students, however, we were never told the significant scientific difficulties that accompanied this teaching.

First, rather than proving macroevolution, this teaching only proved that a shift in the proportion of moths with an advantageous trait over moths with a less advantageous trait can occur; otherwise known as variation within an animal kind—microevolution. This is exactly what the creationist believes as well as what the Bible teaches. In Genesis 30:31-33 we read that although the sheep was one kind of animal, there was variation among this kind. All were not the same color. We read that some were speckled, and some were spotted, and some were brown. Consider the following scenario. If the spots on these sheep were light in color and a predator were on the prowl by moonlight, which sheep would likely be the most vulnerable to attack? Obviously, the ones with the light spots would be more vulnerable since they would stand out better against a dark background. Over a period of time, the proportion of brown sheep to white sheep would certainly increase. This would in no sense prove that macroevolution had taken place anymore than does a population shift in the proportion of dark to light colored moths or light colored people to dark colored people.

Secondly, for decades forthright evolutionists have come out in agreement with this premise and have denied that the change in the proportion of peppered moths proves macroevolution. Over thirty years ago, the renowned evolutionist L. Harrison Matthews wrote in the foreword for the 1971 edition of *The Origin of Species* that "The peppered moth experiments" show "survival of the fittest. But they do not show evolution in progress" (Darwin, 1971, p. ix). Third of all, recent studies have found discrepancies in the original peppered moth studies of the 1950s in contrast to those that took place over the next two decades. The latter studies found almost a reversed trend in the following areas: (A) Dark moths in many unpolluted areas increased in proportion just as those in polluted areas had; (B) Dark moths continued to increase in proportion after pollution controls were in place

The peppered moth, which is fraudulently used to support Darwinian evolution. (Photo by Roy Cripps. Used with permission.)

and light colored camouflage returned to the trees; and, (C) In one area, dark colored moths began decreasing in proportion before the light colored camouflage returned to the trees (Wells, 2000, pp. 144, 145).

One last item conveniently omitted in our public school lessons was that the peppered moth experiment was based on *false assumptions* and *staged photographs*. It turns out that peppered moths never truly rested on tree trunks where they could be camouflaged, but actually hide on the bottom of horizontal branches, high in tree tops where they are protected. In twenty-five years, one group of scientists studying the peppered moth could find only one peppered moth resting on a tree trunk (*Ibid.*, p. 149). It has now been proven that to create the photos, peppered moths were pinned and glued to the bark of trees. This was admitted by a University of Massachusetts biologist, Dr. Theodore Sargent, who conceded to helping glue peppered moths to a tree for a documentary by the pro-evolution PBS series, NOVA (Witham, 1999, p. 28; Coyne, 1998, pp. 35, 36).

Further, it has been shown that the filming of the original peppered moth experiment conducted by H.B. Kettlewell in the 1950s was staged. In order to film footage of birds eating the light colored moths from trees, Kettlewell took cage-bred moths and placed them on tree trunks. At times the moths were so lethargic that Dr. Kettlewell had to warm them on the hood of his car to make them appear naturally wild (Matthews, 1993, p. D3).

In spite of contradictory facts, the peppered moth myth continues to be promulgated as a classic proof of macroevolution in many schools and universities. One staunch evolutionist and anti-creationist, Jerry Coyne, lamented his dismay admitting that "the prize horse in our stable" had been disproved. He went on to say that discovering that the peppered moth story did not prove macroevolution was like, "my discovery, at the age of six, that it was my father and not Santa who brought the presents of Christmas Eve" (Coyne, 1998, pp. 35, 36). Notwithstanding, a 2011 review of twenty-two biology textbooks found that ten textbooks still contain the disproved peppered moth doctrine as proof of macroevolution (Luskin, 2011).

NOTES

_____
_____
_____
_____
_____
_____

Questions

FILL IN THE BLANKS *Using Scriptures from this lesson*

1. "O full of all _____ and all _____, you son of _____ _____, you enemy of all righteousness, will you not cease _____ the straight ways of the Lord?"

2. "But we have renounced the _____ things of _____, not walking in _____ nor handling the word of God _____, but by manifestation of the truth commending ourselves to every man's conscience in the sight of God."

3. "For such are _____ _____, _____ workers, transforming themselves into apostles of Christ. And no wonder! For _____ himself transforms himself into an _____ of light."

4. "That we should no longer be _____, _____ to and fro and carried about with every _____ of _____, by the _____ of men, in the _____ _____ of _____ _____."

5. "Beware lest anyone _____ you through _____ and _____ _____, according to the _____ of men, according to the basic principles of the _____, and not according to Christ."

6. "Let me pass through all your _____ today, removing from there all the _____ and _____ sheep, and all the _____ ones among the lambs, and the _____ and _____ among the goats; and these shall be my wages."

Yes or No

1. ___ Similarities among various animal *kinds* all stem from a common ancestor.

2. ___ Evolutionists teach that because the human eye and the octopus eye are remarkably similar, they must have had a common ancestor.

3. ___ Some animals have organs that are more similar to those on "distantly related" animals than they are to more "closely related" animals, as assessed by evolution.

4. ___ For organs to be truly "homologous" with a common ancestor through evolution, we must find the same genetic code in the DNA in these organs.

5. ___ Haeckel used fraud, deceit, and trickery in his illustrations of various stages of embryonic development.

6. ___ The human embryo contains gill slits but not a tail-bud at one point in its development.

Short Answer

1. How can the Bible's teaching concerning the trickery of men with the gospel (first paragraph of this lesson) be relevant to the study of evolution? _____
_____

2. Explain the circular reasoning that Darwinists use to decide which similarities in animals are "evolutionarily significant." _____
_____
_____

3. Name a couple of similarities shared by alleged evolutionarily unrelated animals that are not shared by those animals to which they are supposedly more closely related. _____
_____
_____

4. What does the so-called "tail" in a developing human embryo in actuality turn out to be? _____
_____
_____

5. What does the increase of dark colored peppered moths over lightly colored peppered moths, or vice versa, in England actually prove in the creationist's favor? _____
_____
_____

6. What did Dr. Gavin De Beer ask in 1938 and 1971, as well as Dr. Michael Denton in 1985 that is still being asked today? _____
_____
_____

Discussion Question in Preparation for Answering Unbelievers and Critics

You are taking freshman biology in college. As your professor is lecturing on macroevolution, he mockingly derides creationists as being crackpots, "flat earthers," uneducated, and dishonest in their dealings. He then says, "In my seventeen years of teaching, I have never had one creationist student offer proof as to why I shouldn't believe the facts supporting evolution." What will you do? (Note: Don't think that this type of bias isn't displayed on a regular basis in colleges and universities). _____
_____
_____
_____
_____
_____
_____

Lesson 8

Defeated Neo-Darwinian Dogmas (III): Darwin's Finches, Fruit Flies, Vestigial Organs and Alleged Horse Evolution

I. Darwin's Finches

Early in the nineteenth century a young Charles Darwin was taught the erroneous doctrine of the "fixity of *species*" by the church of England. This teaching comes from a misunderstanding of Genesis 1:11, 12, 21, 24 that states, "Then God said, 'Let the earth bring forth grass, the herb that yields seed, and the fruit tree that yields fruit according to its kind, whose seed is in itself, on the earth'; and it was so. And the earth brought forth grass, the herb that yields seed according to its kind, and the tree that yields fruit, whose seed is in itself according to its kind. And God saw that it was good... God created great sea creatures and every living thing that moves, with which the waters abounded, according to their kind, and every winged bird according to its kind. And God saw that it was good... Then God said, 'Let the earth bring forth the living creature according to its kind: cattle and creeping thing and beast of the earth, each according to its kind'; and it was so." From this Scripture, the Anglican Church taught that all species were fixed and could not change, even within their *kind*. The problem with this teaching was that the Bible never taught fixity of species, but fixity of *kind* and variation within a *kind*. Within an animal *kind* there

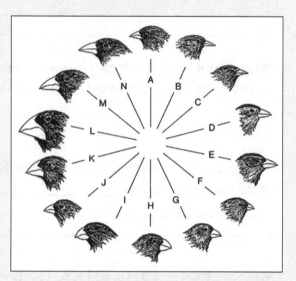

The fourteen species of Darwin's finches that have been discovered in the Galapagos Islands except for the Cocos Island Finch (Finch b). Finch K is the medium ground finch that was studied by the Grants. (Copyright Jody F. Sjogren 2000. Used with permission.)

might be various species, which may produce offspring with varied traits, all within the same *kind*. However, any species within this *kind* will never be capable of changing from one *kind* into another. For instance, there may be various sizes, colors, builds, and temperaments of cats, yet they will always be part of the cat *kind* (see Lesson 3).

Later in his travels, Darwin noticed variation among finches on the Galapagos Islands in the Pacific. Among approximately two dozen islands, fourteen species of these finches, now known as "Darwin's finches," have been discovered. Although the birds are all finches, and are all one kind of animal, there are some slight differences in body structure. The beaks of the birds may be short or long, thick or narrow; their bodies may be larger or smaller than others.

Most of you will recall being taught in school that Charles Darwin's evolutionary doctrine was heavily influenced by these birds. This was never the case. In fact, Darwin wrote only in passing that the finches might have all originated from a

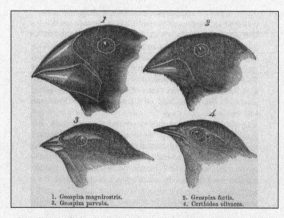

Four of Darwin's finches that he discoverd on the Galapagos Islands, as illustrated by Jonathan Gould, 1845.

Peter and Rosemary Grant went back to the Galapagos Islands in the 1970's and discovered that Darwin's finches were not diversifying into different species, but were interbreeding and may be merging into one species (Photo Credit: University of Nebraska State Museum: explore-evolution.unl. edu/grant.html).

Michael Behe's 2nd book on Intelligent Design, "The Edge of Evolution" clearly shows that microevolution can only proceed up to a point at which it hits a phylogenetic wall that cannot cross into macroevolution and produce a new family of animals.

Lehigh University biochemistry professor Dr. Michael Behe, who met with great acclaim as well as vociferous attacks by macroevolutionists when he published his 2007 book, The Edge of Evolution (Photo credit: brianjones.com)

and becoming much stronger, more healthy, and in the long run might have a better chance of surviving over the traditional species. This type of improvement arises from intermingling of the gene pool, known as hybrid vigor. These facts led many scientists to conclude that over time it is quite possible that the fourteen species of finches might not be evolving into more diverse species; they might all be merging into one! Secondly, scientists have also concluded that the change in beak and body size of the birds due to climate change does not prove long-term evolutionary change. It only serves to prove that the finches oscillate in size from larger to smaller, and then back again over time as one would expect any population of animals to do—variation within a kind. Thirdly, because the finches' mating patterns are based on beak size and songs (learned from parents) and can interbreed with each other, the finches might not be fourteen species after all. They might be just one! (Wells, 2000, p. 172).

Humans may come in different sizes based on the genetic "plasticity" of their DNA, but they will never be anything other than humans. In the photo are Chandra Bahadur Dangi from Nepal (1'9") and Sultan Kosen from Turkey (8'3"). (Photo credit: Peter Macdiarmid/Getty Images)

single species. Not until the 1970s were the finches studied in depth by Peter and Rosemary Grant who supposedly proved the evolution of these birds. On one of the smaller Galapagos Islands, the Grants caught, measured, banded, and released all medium ground finches (finch K). After the drought of 1977, the Grants discovered that 85% of the population of finches died. The birds that were left were slightly larger in body mass and beak width (0.05 mm wider, or ~the thickness of a fingernail). However, after the El Niño rains of 1983, the Grants discovered that the finches reverted to their original smaller size. Over time, then, the size of finches and finch beaks oscillated between larger and smaller, with no appreciable long-term or "evolutionary" change. The researchers also discovered that some of the fourteen species of finches were interbreeding

What we learn from the finches is nothing that we had not already known; principally, that variation exists within an animal kind. Humans come in all shapes, sizes, skin and hair color, with various genetic codes; however, we are all still part of the human kind. Finches are all still part of the bird kind, even as God established in Genesis 1:24. Phylogenetic borders exist to the changes that can occur within an animal kind. A fish kind will never become an amphibian kind will never become a reptile kind will never become a bird kind will never become a mammal kind. And among these groups, a hummingbird will never become an ostrich, a toad will never become a salamander, a snake never will become a crocodile, a goldfish never will become a shark, and an orangutan never will become a human. There is a natural "edge" to the evolution of animals over which genetic variations cannot cross. Each animal is "fixed" within those limits of microevolution. Darwin's finches teach us that a finch is a finch is a finch!

II. Fruit Fly Evolution

In addition to being taught progressive evolution in middle school, in the seventh grade my fellow students and I selectively bred fruit flies (*Drosophila melanogaster*). Our aim was to multiply their mutations and create unusual features in their eyes, wings, and bristles. Mutations are unexpected changes in the DNA (genetic code) of an animal. The experiments worked. We were amazed, and these mutations were and continue to be used as proof that neo-Darwinian evolution occurs. But, can mutations and microevolution in animals and the fruit fly be considered a precursor to macroevolution? The answer is no. For a mutation to be considered significant in macroevolution, it must pass the *PBS Test*. The PBS test requires that mutations be:

Four-winged fruit flies, an inferior mutant offspring, are hindered in their ability to fly and breed. (Photo via The Scientist).

(1) **Persistent.** That is, they must be passed down to their progeny;

(2) **Beneficial.** Secondly the mutation has to be good, helpful, or beneficial to the animal to promote it's survival value, allowing it to out-compete other animals of its kind, and;

(3) **Structural.** The mutation has to be one that will actually change the physical structure of the animal.

Why? Because macroevolutionists that insist one kind of animal (e.g., a cow), will change into another kind of animal (e.g., a whale); hence, great morphological/structural changes must occur. *PBS.* Try to remember that, so the next time someone says there is a mutation that promotes macroevolution you can say, "is it persistent, beneficial, and structural in nature?" At this point in time, no mutations have ever been discovered that impart this type of benefit to any living creature. Some mutations, such as antibiotic resistance in bacteria and pesticide resistance in insects, are touted as an evolutionary advantage. Although these mutations may provide a competitive advantage, at best they only prove variation

within a species, and impart no beneficial structural evolution necessary to validate the neo-Darwinian hypothesis. Further, when a protein is changed in order to resist pesticides or antibiotics, there will be associated negative effects somewhere else in the organism. We call these "pleiotrophic effects." The bacteria may be resistant to the antibiotic, but now it may have a much slower growth rate, and lack the ability to metabolize certain carbohydrates.

The difficulty with relying on mutations to produce evolution is many-fold. (1) Over 99 percent of all mutations are negative and work to the detriment rather than to the benefit of the animal, so says Nobel Prize winner H.J. Muller (1955, pp. 58-68). (2) No PBS mutations have ever been recorded. (3) Even if beneficial mutations could be found, the timetable needed for evolution to occur makes the process impossible. Renowned evolutionist George G. Simpson recognized this problem and admitted that, even if there was a breeding population of 100 million individuals that produced new progeny ever twenty-four hours, the chances of obtaining good evolutionary results from mutations would be once every 274 billion years! (1953, p. 96). However, evolutionists expect us to believe that millions upon millions of evolutionarily beneficial mutations have occurred in transforming "bacteria to Bob" or "microbe to Mary" in their projected time line of no greater than *3.8 billion years*. Remember, in a best case scenario, *one* beneficial mutation would supposedly occur "once every 274 billion years," according to the evolutionists' timeline.

Because the evidence of PBS mutations does not exist, and evolutionists want to cling to their beliefs, they have been increasingly pressured to invent such evidence. Enter the fruit fly. Because the fruit fly can produce a new generation approximately every 10-12 days, selective breeding of the fruit fly can display mutations much quicker than humans that require about twenty years per generation. One scientist, while breeding fruit flies in 1915, noticed a genetic defect in one of the wings. This reproductive line of flies has been preserved ever since. Scientists later

Genetic mutations, such as this five-legged frog, although amusing, produce no lasting benefit as Darwinism might hope. (Photo courtesy of usgs.gov).

This two-headed calf is a good example of a mutated animal that would have no survival value in the wild, thus would die very young.

Nobel prize winners, Drs. C. Nüsslein-Volhard, and E. Wieschaus looked for every possible mutation in the fruitfly, but couldn't find one persistent, beneficial, structural one (Photo credit: Reesfilms.com)

discovered that by breeding this genetic defect (bithorax) to two other mutations (postbithorax, and anterobithorax) they could produce a fly with four wings. However, before jumping to the conclusion that this is evolution in action, consider that this is a harmful rather than helpful mutation. The mutant four-winged fruit fly could never survive in the wild. Unless closely cared for in the laboratory, it will die off there as well. The second set of mutated wings on this fruit fly are non-functional and only serve to hinder the insect from flying or breeding. This is comparable to genetic anomalies of nature that are sometimes seen such as a two-headed or five-legged calf, or the three legged chicken I saw on one occasion. None of these mutations is beneficial and only works toward shortening the life of the animal. However, considering all the selective breeding among fruit flies, at least one PBS mutation should have been found. With regard to the multiple experiments used to induce mutations in fruit flies over the years, notice what the following authors have to say: "...They X-rayed the daylights out of ole *Drosophila melanogaster* [fruit fly]. They changed the eye colors from pink to white and red and black again. They changed the wings this way and that. They worked on the salivary glands. They increased and decreased the number of bristles. They strained and sweated for thousands of hours to change *Drosophila* into something else. What happened? Two things. One, the mutant flies either died over a period of generations, or, they came back to their original, normal conditions!!" (Hall and Hall, 1974, p. 112).

Drs. C. Nüsslein-Volhard, and E. Wieschaus, using a technique called "saturation mutagenesis" searched for every possible mutation involved in the development of the fruit fly and won the 1995 Nobel Prize in medicine for their research. However, they discovered not one mutation that would benefit the fruit fly in the wild (Nüsslein-Volhard and Wieschaus, 1980, pp. 795-801). Other scientists have done the same thing with the zebrafish as well as with a small worm, *C. elegans.* Their results have been the same—no PBS mutations have ever been discovered. Although the mutations produced in the laboratory might make for a good circus side show, they would never be construed as evolutionarily beneficial.

III. Horse Evolution

Certainly you are familiar with the horse diagram of horses supposedly evolving from the early "*Eohippus*" (or *Hyracotherium*) all the way up to the modern day *Equus*. These have been continually used in High School and College text books for more than sixty-five years. The only problem is that for several decades, evolutionists themselves have admitted that these horses did not evolve from one another and lived at the same time. Notice the following quotes from well-known evolutionists of the past and present:

"Prothero and Shubin conclude: 'Throughout the history of horses, the species are well-marked and static over millions of years. At high resolution, the gradualistic picture of horse evolution becomes a complex bush of overlapping, closely related species" (Gould and Eldredge, 1993).

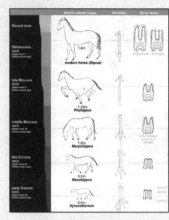

A common textbook representation of alleged horse evolution, which paleontologists now say is inaccurate. If so, then why are these images still used in High School and College text books? (Photo Credit: Mcy Jerry, 2005).

"The family tree of the horse is beautiful and continuous only in the textbooks. . . . The construction of the horse is therefore a very artificial one, since it is put together from non-equivalent parts, and cannot therefore be a continuous transformation series" (Nilsson, 1954).

"The horse is... the classic story of one genus turning into another, turning into another. Now it's becoming apparent that there's an overlap of these genera, and that there were many species belonging to each one. It's a very bushy sort of pattern that is, I think, much more in line with the punctuational model; there isn't just a simple, gradual transition from one horse to another. This is now becoming fairly well-known" (Stanley, 1986).

"The uniform, continuous transformation of *Hyracotherium* into *Equus*, so dear to the hearts of generations of textbook writers never happened in nature" (Simpson, 1953).

IV. Vestigial Organs

The last defeated neo-Darwinian dogma we will examine is the doctrine of vestigial organs. This is the teaching that animals harbor multiple organs that are unneeded vestiges (leftovers or remnants) of organs passed down from our evolutionary ancestors over millions of years. In fact, in 1931 the German Scientist Alfred Wiedersheim, in a book entitled *The Science of Life*, listed 180 useless vestigial organs in the human body alone. Growing up as a child in the public school system, I was taught that the human tailbone, appendix and tonsils were vestiges of our evolutionary past. As recently as 1997, the *Encyclopedia Britannica* had this to say regarding the appendix: "The appendix does not serve any useful purpose as a digestive organ in humans, and it is believed to

be gradually disappearing in the human species over evolutionary time" (p. 491). Let us now examine why the doctrine regarding vestigial organs is the real vestige of neo-Darwinian propaganda.

A. Why no organs can be considered vestigial.

1. The first reason that no organ can be considered vestigial is that there is no way to prove that an organ is useless. Unless one possesses all knowledge of every facet of human physiology, one cannot know for certain that an organ is not utilized by the body for some unknown useful purpose. In fact, all 180 of the organs suggested as being vestigial in the 1931 book have since been shown to be of value!

2. Even if organs were shown to be vestigial, which is impossible, this would in no way *prove* evolution but would illustrate exactly the opposite. Consider the consequences of the idea that 180 organs in the human body at one time served some benefit, but now have become vestigial and are non-functional. The conclusion would be that the human body has become less complex over time rather than more complex. Neo-Darwinian doctrine teaches that information must be added to the genome and structures must be added to the body—not removed!

3. For years physicians in text books have denounced the idea that organs are vestigial. However, the instruction was ignored and evolutionists continued to use the disproved argument in textbooks and public school teachings.

4. Well-studied evolutionists refuse to use this teaching as evidence of evolution today. As far back as thirty-five years ago, one evolutionist in the scientific journal *Evolutionary Theory* denounced this teaching in saying,"...vestigial organs provide no evidence for evolutionary theory" (Scadding, 1981, pp. 173-176).

B. An examination of some supposed vestigial organs.

1. The appendix. What I wasn't taught while in school was that, as far back as the early 1970s, the medical community was teaching the usefulness of the human appendix. Notice the following quote from the 1976 medical textbook *Gastroenterology*: "The appendix is not generally credited with significant function;

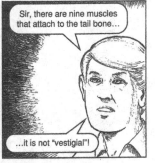

Dr. Keith L. Moore, who, in the medical textbook, Clinically Oriented Anatomy proclaimed the appendix to be a "well-developed lympoid organ" (Photo: Credit: The American Association of Clinical Anatomists famucon.com/clinicalanatomy/news.html).

Cartoon debate between a student and a college professor regarding vestigial organs and evolution. (Copyright 2000 by Jack T. Chick. Reproduced by permission of Chick Publications. www.chick.com).

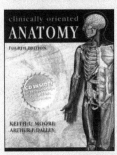
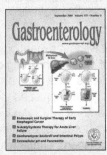
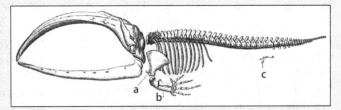

A baleen whale skeleton, with hip bones, proven useful during mating, indicated by "C." Nevertheless, as of February, 2016, Wikipedia still lists the hip bones as vestigial (Photo: Wikicommons, Public Domain).

The scientific journal Gastroenterology, admitted as early as 1976 that the appendix was an organ involved in the immunology of humans. In the textbook Clinically Oriented Anatomy, Keith Moore describes the appendix as "a well-developed lymphoid organ."

however, current evidence tends to involve it in the immunologic mechanism" (McHardy, 1976, p. 1135). Modern medical textbooks go even further in describing the appendix as "a well-developed lymphoid organ" (Moore, 1992, p. 205). As a constituent of the human immune system, the appendix contains lymphatic tissue that helps protect the body against invading microbiological organisms.

2. The tonsils. The tonsils have in the past few decades been shown to work very similarly to the appendix in fighting illnesses in the human body, especially in children. Wikipedia says, "These immunocompetent tissues are the immune system's first line of defense against ingested or inhaled foreign pathogens." Most doctors today, unlike in previous years, do not recommend removing the tonsils of children, except in extreme cases. This is in spite of today's simpler, less invasive methods of removing the organs.

3. The coccyx. The human coccyx, or tail bone, is the last group of vertebrate in the lower spine, fused into a solid column. I was taught in school that this was a vestige organ from our supposed furry-tailed ancestors who roamed about in the trees. The truth is that our tailbone serves many useful purposes in the body. Some of these include being a point of connection for several muscles, providing support for muscles used in the elimination of bodily wastes, and as an aid in sitting and in providing support when standing.

4. Hair on human skin. Evolution says that since man evolved and began to use clothing for warmth, we no longer needed our fur (hair) and thus the muscles at the base of our hair follicles began to recede, and at some point in the future will disappear altogether. For decades, experts have known this to be false. In fact, the hair in human skin serves several useful purposes. One purpose is to compress the sebaceous gland, causing it to excrete oily sebum, moisturizing, waterproofing and protecting the skin. This has been taught in universities for over half a century (Marshall and Lazier, 1946, p. 141), yet the vestigial teaching persists. Another purpose of hair is to keep open the skin follicles which otherwise can become clogged leading to sebaceous cysts.

5. The wings of flightless birds. The wings of flightless birds such as the ostrich, emu, and rhea serve many useful purposes.

Among the functions are (1) providing balance while running, (2) use as a self-cooling heat dissipation system in hot weather, (3) protecting the internal organs in a fall, and in (4) protecting chicks.

6. Supposed "hip bones" in whales. We are told that the longer bones that protrude toward the posterior of the spine in whales were once hips before the mammals evolved to live in the ocean, and that these bones now serve no purpose. It has been known for quite some time that these bones are useful and serve as a point of connection for several muscles, they support the internal organs, and are actually used in mating (Wieland, 1998b, pp. 10-13). Surprise, surprise. In 2014, Dines *et al.* published an article in the journal Evolution which affirmed what creationists have been saying for years; namely, the pelvic bones in whales are not vestigial and are used to position the animals during mating.

The rhea, like the ostrich and emu, have wings that are not vestigial but are useful for at least 4 purposes described in the text (Photo credit: David Long, see Bibliography)

Conclusion

Hopefully the last three lessons have served to expose the error of many neo-Darwinian teachings. We should not expect that men who refuse to submit to divine authority will be above-board in their treatment of the facts. Why should they? Not without exception, this is frequently the rule of order within religious as well as secular circles. "But false prophets also arose among the people, just as there will also be false teachers among you, who will secretly introduce destructive heresies, even denying the Master who bought them, bringing swift destruction upon themselves. Many will follow their sensuality, and because of them the way of the truth will be maligned; and in their greed they will exploit you with false words; their judgment from long ago is not idle, and their destruction is not asleep" (2 Pet 2:1-3, NASB).

NOTES

_____
_____
_____
_____
_____
_____

Questions

TRUE OR FALSE

_____1. There is no observable variation within finches.

_____2. Darwin's finches display variation within an animal kind just as the Bible teaches.

_____3. Mutations in animals lead to macroevolution.

_____4. Observed physical mutations hinder rather than help an animal's chances for survival.

_____5. We must trust all the opinions of scientists because they are the experts.

Fill in the Blank

1. "Then God said, 'Let the earth bring forth the living creature according to its_____: _____ and _____ _____ and _____ of the earth, each according to its _____; and it was so."

2. Over _____ percent of all mutations are _____ and work to the _____ rather than to the _____ of the animal.

3. _____ _____, a process utilized to search for every mutation in the development of an animal, has failed to find one PBS mutation.

4. Some previously considered vestigial organs that are now considered useful include the _____, the _____, the _____, and the _____.

5. "Many will follow their _____, and because of them the way of the _____ will be _____; and in their _____ they will exploit you with _____ words;"

6. For a mutation to be considered evolutionarily significant (or useful) it must meet the _____ test which means it must be _____, _____, and _____.

Short Answer

1. Explain the difference between the early 19th century teaching of the church of England regarding the fixity of species versus the actual biblical teaching of fixity of kinds._____

_____

_____

2. Explain why antibiotic resistance in bacteria and pesticide resistance in insects are not proof of neo-Darwinian evolution._____

_____

_____

3. Name some supposedly vestigial organs. _____

_____

_____

4. How many years did evolutionist Dr. George Gaylord Simpson say would theoretically be necessary for good evolutionary results to occur, and how old do evolutionists believe the world is? _____

_____

_____

5. Provide three reasons why the argument that "vestigial organs prove neo-Darwinian evolution" should not be taught. _____

_____

_____

Discussion Question in Preparation for Answering Unbelievers and Critics

Your homework assignment in science class involves answering questions regarding human anatomy. An essay question is as follows, "Describe what your appendix, tonsils, and pineal gland all have in common and how do these parts of your body provide proof of evolution?" What do you do? _____

_____

_____

_____

_____

_____

_____

_____

Lesson 9

The Geologic Column, the Fossil Record, the Absence of Transitional Fossils, and the Cambrian Explosion

Introduction

The geologic column and the fossil record represent the backbone of evolutionary ideology. Without it, the hypothesis admittedly crumbles. The geologic column must of *necessity* contain irrefutable evidence for the historical account of evolution, or else the doctrine clearly falls under its own weight.

I. Introduction to the Geologic Column

The geologic column is the strata of various layers of rock in the earth's crust. Altogether, the supposed geologic column is 1.5 miles thick, although there is no place on earth where it is represented that way. Does that not strike you as being strange?! Geologists, therefore, put together various rock layers around the world to compose a "theoretical" geologic column. Most evolutionists ascribe these formations to the theory of *uniformitarianism*: the belief that rock layers were created by gradual accumulation of sediment over billions of years. They believe that the rate at which the sediment was laid down in the past is *uniform* and equivalent to the rate at which sedimentation is laid down today. On the other hand, creationists, supported by the biblical account, suggest that the sedimentary and igneous-formed (molten-rock formed) layers were laid down in a rather short period of time (A) during the creation of the earth and (B) during the great Noahic flood (see Gen.

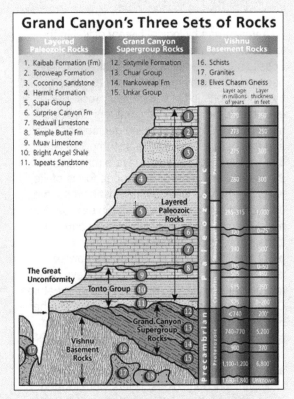

Geologic or stratigraphic layers of the grand canyon with alleged ages according to the evolutionists' subjective timescale. (Photo: U.S. Park Service).

1:6-9; and Gen. chs. 7, 8). Sedimentary rock is produced by the layering of earth matter and the subsequent compaction of those layers into a solid strata in the earth's crust. These geologic layers and sublayers have been assigned ages and names by evolutionary geologists (Table 1).

Geologic Periods	Predicted Age
Recent strata	Present to 1.8 Million Years Old
Cenozoic	5 to 70 Million Years Old
Mesozoic	70 to 200 Million Years Old
Paleozoic (includes the Cambrian Period)	200 to 530 Million Years Old
Proterozoic (Precambriam Age)	530 Million to 1 Billion Years Old
Archeozoic (Precambrian Age)	1 to 1.8 Billion Years Old

Table 1. Geologic periods and ages as estimated by evolutionary-leaning uniformitarian geologists

A Ph.D. creation scientist pictured excavating a Stegosaurus, together with a Camarasaurus, near Dinosaur, Colorado. (Photo courtesy of Don Patton and Steve Rudd. www.bible.ca.)

A very crude biological history of life, in the form of fossils, may be found in these layers. Fossils are the remains of organisms that have been preserved by mineralization or either by impressions left in sedimentary rock. For evolution to be true, three great assumptions must be made concerning the geologic column and the fossil record. These assumptions are that (A) the oldest geologic layers must contain the earliest and most undeveloped forms of life, while (B) the youngest layers must hold the most developed and evolutionarily complex organisms, and (C) amid these layers must exist enumerable transitional fossils or intermediate forms; proof that life evolved in a gradual stepwise fashion from the most simple and undeveloped microorganisms to the most complex of creatures. Creationism and the Bible account, on the other hand, predict that the geologic column should present the following four things. First, (A) we should see a sudden appearance of fully formed animals interspersed between geologic layers. However, since the oldest layers were formed during the creation of the world (Gen. 1:6-9), (B) there should be no life present in these layers. Also, (C) some of the early layers may contain an abundance of smaller aquatic and less mobile life forms unable to escape to higher ground during the first stages of the great deluge of Noah's day. (This topic will be considered more extensively in Lesson 11.) And fourth, (D) according to the Bible, no transitional or intermediate fossils between animal *kinds* should be found. All animals in the fossil record should be fully formed and clearly distinct *kinds*, as denoted in the Genesis account (Gen. 1:21, 24, 25). Bear these four distinctions in mind as you finish this lesson.

II. The Fossil Record: The Evolutionists Only Hope

If the supposed evolution of life occurred, it may only be found in the fossils. If it can't be found there, then it can't be found anywhere, and the myth of evolution is shattered. Darwinists admit this fact: "We must look to the fossil record for the ultimate documentation of large-scale change. In the absence of the fossil record, the credibility of evolutionists would be severely weakened. We might wonder whether the doctrine of evolution would qualify as anything more than an outrageous hypothesis" (Steven Stanley, 1979). "That evolution actually did occur can only be scientifically established by the discovery of the fossilized remains of representative samples of those intermediate types… " (Clark, 1955, p. 7). "…Naturalists must remember that the process of evolution is revealed only through fossil forms" (Grassé, 1977, p. 4). "The number of intermediate and transitional links between all living and extinct species must have been inconceivably great" (Darwin, 1872, p. 266). "The most important evidence for the theory of evolution is that obtained from the study of palaeontology… it was the discovery of various fossils and their correct placing in relative strata and age that provided the main factual basis for the modern

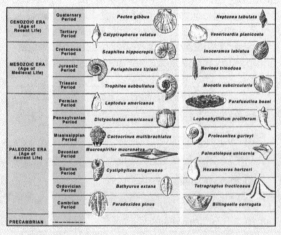

A chart of index fossils published by the U.S. Geologic survey to determine the ages of geological layers.

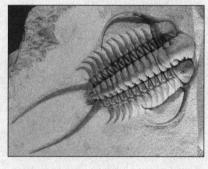

A trilobite (Cheirurus ingricus species) discovered in Russia in 1845. This could be used as an index fossil for geologic layers supposedly 390 - 500 million years old (Photo: Smithsonian Insitute).

THEOPHILUS   *Dating Technique*

This fossil is 3 billion years old.

How do you know?

It was in this rock strata.

Oh! Well, how can you be certain of the age of this rock strata?

This 3 billion year old fossil was in it!

?

©BOB WEST 1997
www.theophilus.org

Rights owned and administered by Bob West. Used with permission.

view of evolution" (Kerkut, 1960, p. 134). Unless evolution can be found in the fossils, it cannot be proved.

III. A Word on the Dating of Fossils and the Geologic Column: *"Heads I Win, Tails You Lose"*

Although radiometric dating of archaeological artifacts and rocks is a very popular topic, a topic addressed in a previous lesson, this is not the primary tool that geologists have historically used to assign dates to the geologic layers. The most common method of dating geologic layers is by using "index fossils"—fossils that have been assigned ages based on when Darwinists believe they *should have evolved*. For example, if an evolutionist uncovers an ancient organism such as a trilobite, he might assign to it an age of 530 million years. But, there is no scientific *proof* that the trilobite is actually 530 million years old. So, when questioned as to how he can ascertain the age of the fossil he says, "Because it was found in Cambrian rock that is 530 million years old." But when asked how he knows that this Cambrian rock is 530 million years old he might disdainfully reply, "Because we found a 530 million year old trilobite index

fossil in it!" And thus the circular reasoning continues where the Darwinist has devised an unscientific, fool-proof system so as never to be proven wrong. *"Heads I win, tails you lose."* One Victorian philosopher, albeit an evolutionist and devout atheist, took issue with this unscrupulous and dishonorable practice among the scientific community in his 1859 essay entitled, "Illogical Geology." Herbert Spencer derided the unprincipled technique of using index fossils to date geologic strata. He was disheartened with the dishonesty of a system that often turned up fossils in the *"wrong strata,"* invalidating the method entirely (1966, pp. 192-210).Nevertheless, the practice continues today.

IV. The Facts of the Geologic Column and the Fossil Record

A. Geologic Layers Do Not Reveal Successionary Evolution. The evolutionist predicts that the oldest layers of rock must contain the earliest and most undeveloped forms of life, while the youngest layers must contain the most developed and evolutionarily evolved organisms. Is this what they find? Not at all. One scientific journal reported, "Since the early days of the acceptance of the standard geologic column, fossils have been turning up in 'wrong' places as more and more fossils have been collected, and this process continues to this very day" (Woodmorappe, 2000, pp. 110-116). Almost the entirety of the geologic column (from the Cambrian layer on up) contains oceanic fossils, indicating a global flood. In many places, these ocean fossils are intermixed with fossils of land-dwelling animals. Discussing the "top of the Cretaceous" layer, evolutionist Dr. Derek Ager says, "It is, in fact, very difficult to correlate the marine and the non-marine strata at this level" (1993b, p. 190). The following are a few of the myriad examples of life forms discovered in geologic strata *older* than is possible, according to evolution. In other words, according to evolution, the following life forms supposedly evolved millions of years *after* the rock layers they are found in were formed. The list includes Dasycladalean algae, pipiscids (metazoan animals lately found in the Cambrian strata), agnathan (jawless) fish, *Lystrosaurus* reptile, the *Neoguadalupia* sponge, the *Camptochlamys* bivalve, and the *Parafusus*

One of the only life forms found in the precambrian layers are these algal mats called "Stromatolites." So, where are all the transitional fossils that predate the forty animal body plans found in the Cambrian layer?

Fossilized stromatolite columns near Laterrière, Quebec. Stromatolites are microbial mats formed in shallow waters by bacteria producing biofilms, and are some of the only forms of life detected in the precambrian strata, (Photo: André-P. Drapeau P., Wikicommons).

Modern stromatolites in Shark Bay, Western Australia (Photo: Harrison, 2005, Wikicommons).

Modern day Stromatolites, which are algal mats, also found fossilized in the precambrian layers.

gastropod (*Ibid.*). Time fails us to discuss each organism, but suffice it to say these life forms were supposed to have evolved long *after* the ancient geologic strata they were found in were formed. In fact, in 1992, J.J. Sepkoski, a professed evolutionist, reported that in only ten-year's time, 1026 fossil families underwent an *increase* in their stratigraphic range. That is to say, 1026 fossil families from 1982-1992 were discovered to be much older than previously thought and often times much older than is evolutionarily possible (p. 7). Other out-of-place fossils include a small dinosaur discovered in the stomach of a dog-like mammal (Hu, *et al.*, 2005)

B. Transitional (intermediate) forms: Darwin's Achilles heel. Evolution teaches that amidst geologic layers, transitional fossils, or intermediate forms must be found: proof that animals evolved in a gradual stepwise fashion from the most simple to the most complex. So, where are they? Where are the millions of transitional fossils between each of the major groups of animals? They simply do not exist! If you are in doubt, hear from the evolutionists themselves. Charles Darwin in *The Origin of Species* admitted that these transitional forms could not be found, laying waste to his theory: "Why is not every geologic formation and every stratum full of such intermediate links? Geology assuredly does not reveal any such finely-graduated organic chain; and this is the most obvious and serious objection which can be urged against the theory" (1872, pp. 264-265).

In more than 150 years since Darwin's writing, you would think that these intermediate links would have been found, right? Wrong. Notice more recent comments by the evolutionary community: "Paleontologists are traditionally famous (or infamous) for reconstructing whole animals from the debris of death. Mostly they cheat" (Bengston, 1990).

"Instead of filling in the gaps in the fossil record with so-called missing links, most paleontologists found themselves facing a situation in which there were only gaps in the fossil record, with no evidence of transformational evolutionary intermediates between documented fossil species. Without fossil intermediates to back up Darwinian predictions of how evolution works, the turn of the century saw both paleontology (an evolutionary discipline) and gradual change via natural selection (an evolutionary model) fall on hard times" (Schwartz, 1999).

"The expected Darwinian pattern of a deep fossil history of the bilaterians, potentially showing their gradual development, stretching hundreds of millions of years into the Precambrian, has singularly failed to materialize" (Budd and Jensen, 2003).

"The oldest known snakes in the fossil record have no 'intermediate form' characteristics, and they are no different from snakes of today" (Carroll, 1990).

"The evolution of the turtle is one of the most remarkable in the history of the vertebrates. Unfortunately the origin of this highly successful order is obscured by the lack of early fossils, although turtles leave more and better fossil remains than do other vertebrates. . . Turtles allegedly sprang from the 'primitive' reptiles called cotylosaurs, yet intermediates are 'completely lacking' " (Encyclopedia Brittanica, 1992).

"Although an almost incomprehensible number of species inhabit Earth today, they do not form a continuous spectrum of barely distinguishable intermediates. Instead, nearly all species can be recognized as belonging to a relatively limited number of clearly distinct major groups, with very few illustrating intermediate structures or ways of life" (Carroll, 1997).

"Given the fact of evolution, one would expect the fossils to document a gradual steady change from one ancestral form to the descendants. But this is not what the paleontologist finds. Instead, he or she finds gaps in just about every phyletic series. . . This raises a puzzling question: Why does the fossil record fail to reflect the gradual change one would expect from evolution?" (Mayr, 2001).

"The earliest fossils of Homo, Homo rudolfensis, and Homo erectus, are separated from Australopithecus by a large, unbridged gap. How can we explain this seeming saltation? Not having any fossils that can serve as missing links, we have to fall back on the time-honored method of historical science, the construction of a historical narrative" (Mayr, 2004).

Dr. Jeffrey Schwartz says that we, "are still in the dark about the origin of most major groups of organisms. They appear in the fossil record as Athena did from the head of Zeus—full-blown and raring to go, in contradiction to Darwin's depiction of evolution as resulting from the gradual accumulation of countless infinitesimally minute variations" (Schwartz, 1999, p. 3).

"Explained on an intelligent design t-shirt., 'Fact: Forty phyla of complex animals suddenly appear in the fossil record, no forerunners, no transitional forms leading to them; 'a major mystery,' a 'challenge.' The Theory of Evolution —exploded again (idofcourse.com).' Although we would dispute the numbers, and aside from the last line, there is not much here that we would disagree with" (Peterson et al., 2009).

"Darwin recognized that phyletic gradualism is not often revealed by the fossil record. Studies conducted since Darwin's time likewise have failed to produce the continuous series of fossils predicted by phyletic gradualism" (Hickman et al., 2000).

"We still have no evidence of the nature of the transition between cephalochordates and craniates. The earliest adequately known vertebrates already exhibit all the definitive features of craniates that we can expect to have preserved in fossils. No fossils are known that document the origin of jawed vertebrates" (Carroll, 1997, p. 296).

"The idea that one can go to the fossil record and expect to empirically recover an ancestor-descendant sequence, be it of species, genera, families, or whatever, has been and continues to be, a pernicious illusion" (Nelson, 2004).

"No fossil is buried with its birth certificate. That, and the scarcity of fossils, means that it is

The 2010 article in the journal Nature that dethroned Tiktaalik as a transitional species between fish and tetrapods (Niedźwiedzki).

Tiktaalik was once considered a transitional fossil from fish to tetrapods [animals that walk on all fours]. Seven years ago an article was published in the journal Nature entitled, "Tetrapod trackways from the early Middle Devonian period of Poland" (Niedźwiedzki, 2010). The authors reveal tetrapod tracks that predate the oldest tetrapod remains by a supposed 18 million years (Myr) and predate the earliest elpistostegalian fishes by about 10 Myr. In other words, 10 Myr before Tiktaalik existed, there were already tetrapods roaming the planet. So Tiktaalik is not a missing link (transition/intermediate) (Photo: Deretsky, 2007).

Dr. Stephen C. Meyer, author of "Darwin's Doubt" and "Signature in the Cell" is one of the foremost leaders of the modern intelligent design movement and is a Fellow of the Discovery Institute. He is also one of the most targeted ID experts, by the hostile evolutionist crowd.

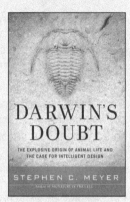

Stephen Meyer's New York Times best selling book, which exposes the failings of modern evolutionary ideas based on a lack of transitional/intermediate fossils prior to the "Cambrian Explosion."

effectively impossible to link fossils into chains of cause and effect in any valid way. . . To take a line of fossils and claim that they represent a lineage is not a scientific hypothesis that can be tested, but an assertion that carries the same validity as a bedtime story–amusing perhaps even instructive, but not scientific" (Gee, 1999).

"Many people suppose that phylogeny can be discovered directly from the fossil record by studying a graded series of old to young fossils and by discovering ancestors, but this is not true... There is no way of knowing whether a fossil is a direct ancestor of a more recent species or represents a related line of descent (lineage) that simply became extinct" (Knox *et al.*, 1994).

"The absence of fossil evidence for intermediary stages between major transitions in organic design... has been a persistent and nagging problem for gradualistic accounts of evolution... I regard the failure to find a clear 'vector of progress' in life's history as the most puzzling fact of the fossil record" (Gould, 1982, p. 140; 1980, p. 127).

As an illustration, for evolutionists to have you believe their hypothesis, despite the absence of transitional fossils, is akin to your stock broker talking you into giving him ten thousand dollars to invest in an alleged fund that he has heard of, but has never seen proof of and doesn't know if it even exists. Would any clear-thinking individual take part in such a scheme? Of course not. Then why

would one invest his faith of the origin of life in the evolutionist who has just as little proof to offer as the stock broker? Notice one more quote from Dr. Stephen J. Gould the most prolific evolutionary of the twentieth century, "The extreme rarity of transitional forms in the fossil record persists as the trade secret of paleontology, the evolutionary trees that adorn our textbooks have data only at the tips and nodes of their branches: the rest is inference, however reasonable, not the evidence of the fossils" (1977, pp. 12-16). What does Dr. Gould mean when he says that the evolutionary trees only have data at the nodes of the branches? He is explaining that all known fossils appear fully formed with none of the transitional linking "branches" that should appear lower in the column, from which these organisms would have evolved. So where are they? Where are the millions of transitional fossils? They don't exist.

C. The sudden explosion of life—the "Cambrian Explosion." If evolution had occurred, we would find an innumerable host of transitional fossils. However, just the opposite is found. Fossils always reveal that simple as well as complex animals appear suddenly and without evolutionary intermediate forms, which some Darwinists call the "Cambrian Explosion." In the Cambrian layer of the geologic column, there is a surprising and sudden appearance of a diversity of animal life. Every major animal body plan (approximately forty different animal body plans), including vertebrates, appears, unexpectedly to the evolutionist, in the "Cambrian Explosion." The most serious problem for the evolutionist is the lack of any precambrian transitions or intermediates predating animals in the Cambrian. For years evolutionists expected to find the transitional, intermediate missing link fossils in the Precambrian layer, but mostly all they find there is a type of algal mat called a stromatolite (see previous pictures) and possibly bacteria.

Some Evolutionists Admit the Inexplicable Sudden Appearance of Animal Kinds in the Fossil Record

A world famous paleontologist and a geologist state, "There are numerous

Artistic reconstruction of Haikouichthys. This was one species of fish that suddenly appears in the Cambrian explosion, with no missing link transitional ancestors that predate its existence (Talifero, 2011).

reports of older trace fossils, but most can be immediately shown to represent either inorganic sedimentary structures or metaphytes [land plants], or alternatively they have been misdated" (Budd and Jensen, 2000). Richard Monastersky, news editor for the journal *Nature,* said, "Before the Cambrian period, almost all life was microscopic, except for some enigmatic soft-bodied organisms. At the start of the Cambrian, about 544 million years ago, animals burst forth in a rash of evolutionary activity never since equaled. . . Paleontologists have proposed many theories to explain this revolution but have agreed on none" (1997). Evolutionist Dr. Derek Ager states, "I always wonder why palaeontologists pay so much attention to sudden extinctions and so little to sudden appearances, such as the spiny foram Hantkennina in the Eocene and (my own favourite) the distinctive brachiopod Pergrinella in the Early Cretaceous. These have no apparent ancestors, but appear 'out of the blue' in many parts of the world with no 'mothers or fathers' . . .It now seems that we do not often see the gradual emergence of a new species either. All these things are sudden and, in a sense, catastrophic" (Ager, 1993b, pp. 189-196).

Dr. Gould further explained, "Most species exhibit no directional change during their tenure on earth. They appear in the fossil record looking much the same as when they disappear... a species does not rise gradually by the steady transformation of its ancestors; it appears all at once and 'fully formed' " (Gould, 1977, pp. 12-16).

Take our example of the bat, pterodactyl, and bird. All of these animals appear fully developed, fully winged, and fully flying animals in the fossils. The oldest bat fossil (assigned

the geologic age of ~50 million years) was found in the Messel Oil Pit outside Darmstadt, Germany. What does it look like? It is a fully developed flyer with a sonar inner ear construction identical to modern bats. The oldest fossilized bat is structurally identical to today's bat, undergoing no evolutionary change in the evolutionist's supposed fifty million year span of time. Evolutionist Dr. Paul Sereno disappointingly stated, "For use in understanding the evolution of vertebrate flight, the early record of pterosaurs and bats is disappointing: Their most primitive representatives are fully transformed as capable fliers" (1999, p. 2143). Why might the appearance of these fully developed animals in the fossil record be described as an "explosion" by some? Because in the evolutionist's scheme of things, each fully organized biological creature must have taken millions of years to develop all of its complex organs and distinguishing biological characteristics. However, when one peers into the earth, this is not what is found. It is as if all ancient life forms in the fossil record cry aloud, "I arrived on this earth suddenly, abruptly and rapidly; not slowly, gradually, progressively or evolutionarily." If the fossils could speak, by their appearance, it is almost as if they would say, "I was created." That is the conclusion God says mankind should reach from studying his creation (Ps. 19:1-3; Rom. 1:19-20).

V. What about the Alleged Discovery of Transitional Fossils?

Very few discoveries have ever been fully agreed upon by evolutionists as representing intermediate links to ancestral animals. Among the few transitional fossils that have been suggested, all are highly dubious and questionable at best, or utterly erroneous and/or falsified at worse. Let's look at the supposed transitional form that has been heralded most frequently in public schools.

Archaeopteryx. The extinct fossilized bird known as the *Archaeopteryx* has for years been paraded as the "missing link" of reptile-bird evolution. For decades grade schools, high schools, and universities have adorned their textbooks with the *Archaeopteryx* as the premiere illustration of evolutionary transition. As late as 1998, the National Academy of Sciences wrote, published, and distributed a book to public school teachers encouraging them to teach evolution. The book entitled, *Teaching about Evolution and the Nature of Science*, uses *Archaeopteryx* as a subject example and says, "It's a fossil that has feathers like a bird but the skeleton of a small dinosaur. It's one of those missing links that's not missing anymore" (1998, p. 8). The use of this example is especially misleading considering the fact that the majority of Darwinists no longer believe that *Archaeopteryx* is an example of evolutionary transition and consider it fully, 100% bird. For starters, the *Archaeopteryx* has almost identical features to a South American bird, the hoatzin. Evolutionists previously claimed that the claws on the end of *Archaeopteryx's* wings indicated a transition between reptiles and birds. However, birds today have the same feature (*e.g.*, the hoatzin, the African touraco, and the ostrich all have claws at the end of their wings). Next, science has shown that *Archaeopteryx's* plumage was not composed of half scales, half feathers. Instead, the bird was fully feathered. Additionally, geologists have discovered entirely developed birds (those that supposedly evolved from *Archaeopteryx*) in geologic layers much older than those that *Archaeopteryx* have been found in (Lyons, 2003). The verdict is in. The *Archaeopteryx* was not a transitional animal; it was a bird, a whole bird, and nothing but a bird. If you won't accept the word of the creationists, then how about three leading evolutionists who reject the *Archaeopteryx* reptile/bird myth? "Paleontologists have tried to turn *Archaeopteryx* into an earth-bound, feathered dinosaur. But it's not. It is a bird, a perching

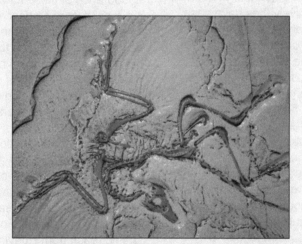

Cast of a fossilized Archaeopteryx, shown to be an extinct bird. No missing link here. (Photo by Joshua Gurtler at the Smithsonian National Museum of Natural History).

bird. And no amount of 'paleobabble' is going to change that" (Feduccia, 1993).

"*Archaeopteryx* probably cannot tell us much about the early origins of feathers and flight in true protobirds because *Archaeopteryx* was, in a modern sense, a bird" (Feduccia, 1993).

"The most striking feature of Archaeopteryx is its well-developed feathered wings. These wings are not significantly different in size and shape from those of modern birds such as magpies or coucals, and they give every indication that *Archaeopteryx* was a flying bird. The feathers also appear to be strong evidence of flight ability . . . . In *Archaeopteryx* the feathers are remarkably similar to those of modern birds" (Rayner and Wooten, 1992).

Birds found earlier in the fossil record than the theropod dinosaurs from which they supposedly evolved.

Further, today there is a whole movement of evolutionists who *deny* that birds evolved from dinosaurs. One of these is John Reuben, Zoologist at Oregon State Univ., who said, "For one thing, birds are found earlier in the fossil record than the dinosaurs they are supposed to have descended from. That's a pretty serious problem, and there are other inconsistencies with the bird-from-dinosaur theories. . . A velociraptor did not just sprout feathers at some point and fly off into the sunset. . . It just seems pretty clear now that birds were evolving all along on their own and did not descend directly from the theropod dinosaurs, which lived many millions of years later. . . Frankly, there's a lot of museum politics involved in this, a lot of careers committed to a particular point of view even if new scientific evidence raises questions" (OR St. Univ., 2009). Vertebrate paleontologist Dr. Robert A. Martin, stated that the theropods "all

The African Tauraco has claws on the ends of its wings, similar to those on Archaeopteryx (Photo: KrisMaes, 2012).

The South American Hoatzin has claws on the ends of its wings, similar to those on Archaeopteryx (Photo: Kate, 2011).

occur in the fossil record after *Archaeopteryx* and so cannot be directly ancestral" (Martin, 2004). Also see Swisher *et al.* (1999).

Conclusion

What does the geologic column display? Primitive creatures in primitive rock, and more developed creatures in younger rock? Transitional fossils of animals representing intermediate links in the chain of evolution? No. Instead, we see examples of simple as well as complex organisms scattered throughout these layers. We witness an "explosion" of creatures fully developed in the Cambrian layer with no hint of evolutionary workings. One evolutionist admits this much in stating, "In any case, no real evolutionist, whether gradualist or punctuationist, uses the fossil record as evidence in favor of the theory of evolution as opposed to special creation... " (Ridley, 1981, p. 831). If the fossil record is the evolutionist's last hope, where does that leave him? This subject will be considered in the following lesson.

NOTES
_____
_____
_____
_____
_____
_____

Questions
MATCHING

_____ 1. The fossil bird, once heralded as the missing link.

_____ 2. Rock that is formed by accumulation of debris.

_____ 3. The evolutionists' "only hope."

_____ 4. How the geologic column may have formed.

_____ 5. The sudden appearance of fossilized animals in the geologic column.

_____ 6. *"Heads I win, tails you lose."*

A. The Fossil Record

B. Sedimentary Rock

C. The Noahic Flood

D. *Archaeopteryx*

E. Circular reasoning in the dating of geologic strata and fossils found therein

F. The Cambrian explosion

True or False

_____ 1. Uniformitarianism is the belief that the majority of sedimentary rock layers have always accumulated at a uniform rate.

_____ 2. As Darwin predicted, many transitional links have been discovered.

_____ 3. The Paleozoic age *was* 200 to 530 million years ago.

_____ 4. Fossils and the geologic column are only dated by radiometric methods.

_____ 5. In ten years time, over a thousand fossils were proven to be older than was previously thought.

_____ 6. Transitional fossils, intermediate fossils, transitional forms, and missing links all mean the same thing.

Short Answer

1. Explain why evolutionists believe that the oldest rock must contain the least developed of organisms and why younger rock must contain the most complex of organisms. _____

_____

_____

2. Why should Genesis 1:6-9 have any bearing on fossilized remains in the geologic column? __

_____

3. According to Genesis 1:21, 24, 25, why should there be no transitional fossils in the geologic column?_____

_____

4. Why is the fossil record considered the evolutionists' only hope? Can't evolution be proven in other ways? _____

_____

5. Does the creation "cry out" to us? If so, how? (See Ps. 19:1-3 and Rom. 1:19-20.) _____
_____
_____

6. Explain what the evolutionist wants to find in the geologic column versus what the Bible
suggests should be found there. _____
_____
_____

Discussion Question in Preparation for Answering Unbelievers and Critics

You are hiking with your brother or sister and some of their friends. As you round the side of a small hill the group notices thirty or forty sedimentary layers in the side of a rocky bank. One individual in the group dislodges two rocks, one from a higher layer and one from a lower layer. He holds these stones out at arm's length and comments, "Isn't it amazing that this rock I'm holding in my right hand is millions of years older than the rock I'm holding in my left hand?" What do you say? _____
_____
_____
_____
_____
_____

Lesson 10

"Hopeful Monsters," A New Hope?
(The Doctrine of Punctuated Equilibrium)

Introduction

Recall from Lesson nine that many of the most studied neo-Darwinian evolutionists have cast serious doubt on the proposed mechanism of evolution. They have arrived at this conclusion because of the absolute lack of any intermediate (transitional) fossils that would indicate evolutionary change. This is in spite of the ~300,000 fossil species that have been unearthed in the past 150 years. Also, recall from chapter three that over twenty-three new mechanisms for macroevolution have been proposed, although evolutionists cannot agree on what should be added to or replace Neo-Darwinism. One of the more famous proposed alternative mechanisms known as "Punctuated Equilibrium" will be the subject of this chapter.

I. Admitted Neo-Darwinian Difficulties Led to the "Hopeful Monster" Position

Numerous difficulties associated with Darwin's theory of gradual evolution were discovered soon after his 1859 publication. Oddly enough, the loudest opponents were often from the evolutionist rather than from the creationist camp. These detractors expressed

their chagrin with the idea of slow, geologically recorded change. One distinguished evolutionist, Dr. F. Ayala, is reported as concluding, "I am now convinced from what the paleontologists say that small changes do not accumulate" (as quoted in Lewin, 1980, pp. 883-887). This is the problem. Evolution insists that over *millions* of years, small mutational changes accumulated into new kinds of animals. Two other renowned

A fossilized trilobite, which over supposedly millions of years experienced no change in the structure of their highly complex eyes, leading Dr. Niles Eldredge to eventually believe in an explanation called punctuated equilibrium.

Anthropologist, Dr. Ernest Hooten, recognized as early as 1937 that mutations could not account for macroevolution (Photo: Wikicommons).

Ian Tattersall and fellow curator of the American Museum of Natural History with Dr. Niles Eldredge. Both agreed in their 1982 publication that the fossil record did not reveal gradual evolutionary change.

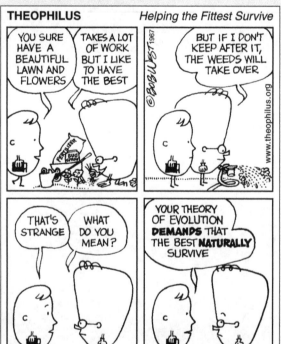

THEOPHILUS   Helping the Fittest Survive

YOU SURE HAVE A BEAUTIFUL LAWN AND FLOWERS

TAKES A LOT OF WORK BUT I LIKE TO HAVE THE BEST

BUT IF I DON'T KEEP AFTER IT, THE WEEDS WILL TAKE OVER

THAT'S STRANGE

WHAT DO YOU MEAN?

YOUR THEORY OF EVOLUTION **DEMANDS** THAT THE BEST **NATURALLY** SURVIVE

Renowned advocates of the "hopeful monster," "quantum evolution," and "punctuated equilibrium" views. Left to right: Drs. O.H. Schindewolf, R.B. Goldschmidt, G.G. Simpson, S.J. Gould, and N. Eldredge (Schindewolf photograph from the Yearbook of the Academy of the Sciences and the Literature, 1971 [p. 75], Mainz, Germany. Goldschmidt photograph from the Alan Cock collection. Used with permission. G.G. Simpson photograph public domain. S.J. Gould illustration by Rob Baker via Apologetics Press. Used with permission. N. Eldredge photograph provided by the American Museum of Natural History. Used with permission.

evolutionists noted, "If ever there was a myth, it is that evolution is a process of constant change" (Eldredge and Tattersall, 1982, p. 8). Celebrated evolutionist Dr. Ernest Hooton declared, "Many anthropologists (including myself) have sinned against genetic science and are leaning upon a broken reed when we depend upon mutations" (1937, p. 118). As you can see, the failings of mutations were known as early as 1937. Fast-forward forty years and hear from evolutionist Dr. Pierre-Paul Grassé: "Today our duty is to destroy the myth of evolution, considered as a simple, understood, and explained phenomenon which keeps rapidly unfolding before us." He then explained why evolutionists continue teaching a theory they know to be false: "The deceit is sometimes unconscious, but not always... people... purposely overlook reality and refuse to acknowledge the inadequacies and falsity of their beliefs" (Grassé, 1977, p. 8). That is *not* to say that these gentlemen no longer believe in evolution. Far from it, they are simply heralding the fact that the neo-Darwinian *mechanism* by which evolution transpired is inherently flawed. Nevertheless, not one of them could provide a sound mechanistic alternative. As we illustrated in lesson three, once individuals reject the God of the Bible and the Genesis creation account, they have painted themselves into a corner of believing a naturalistic view concerning the origin of life.

II. What Has Happened to These Folks?

How can the supposed scientific and intellectual giants of our time fail to perceive the mountains of evidence for intelligent design staring them in the face, and opt for an un-

grounded belief for which many admit there is no evidence? From a spiritual perspective, we know what has happened. When people are taught any form of error long enough and loudly enough, they will eventually believe it as they become hardened against the truth. God warns of such hardening of the heart and blinding of the mind (see Prov. 28:14; 2 Cor. 3:14-16; 4:3, 4; Eph. 4:17-19; Heb. 3:13; 1 John 2:11). After this petrification occurs, an individual may become so calloused as to believe almost any lie, regardless of how clear the facts to the contrary may be (Rom. 1:24-25).

III. "Hopeful Monsters"—the Other Evolution

As the story goes, some evolutionists, being a little more scrupulous than others, have confessed to the unreliability of classical neo-Darwinian evolution of gradual, mutation-accumulating transformation that is not supported in the fossil record. So, what were they to do now that the fossils, their last hope, provided no backing for their hypothesis? In a strange turn of events, individuals recognizing this fact postulated an alternative theory. In 1940, the famous German geneticist, and later U.C. Berkeley professor, Dr. R.B. Goldschmidt, followed in the footsteps of Europe's one time top paleontologist, Dr. O.H. Schindewolf, proposing the idea of what he called "hopeful monsters." His belief was that instead of evolution occurring slowly over millions of years, which he understood was unsupported in the fossil record, evolution must have taken place all at once. It must have occurred in large macromutational leaps such as in a bird hatching from a reptile egg through *macro* rather than *micro* mutations. The animal

Microbiologist and biophysicist Professor Carl Woese from the Univ. of Illinois and co-authors in a 2008 publication (see Roberts et al). proposed "saltationism" or "saltation evolution," which is the idea that an animal can transform species in one generation. (Photo: Hamerman, 2004).

would be hailed a "monster" because it would have to be a monstrously mutated creature and "hopeful" because, unlike over 99.9% of drastically mutated animals, its mutations would not be lethal, thus permitting it to procreate. One writer, explaining Goldschmidt's dilemma, said that he, "observed, after forty years working with micromutations, that they seemed to lead nowhere... Goldschmidt was puzzled: if mutations lead nowhere... Where did teeth, blood circulation, compound eyes and the poison apparatus of the snake come from? Logically enough, he broke with Darwinian graduation, suggesting that 'mega' or 'macro' mutations must provide the answers" (Pitman, 1984, p. 72). Although his idea was patently absurd, Dr. Goldschmidt was honest enough to reject the Darwinian mechanism of gradual evolution while refusing to reject the theory in its entirety. Goldschmidt's new idea, as you can guess, was laughed to scorn in the evolutionary community, although none of his detractors could provide any practicable alternatives. A similar, but modified version of the "hopeful monster" mechanism is still held by some and is called "saltationism" or "saltation evolution," which is the idea that an animal can transform species very rapidly. Carl Woese and co-authors proposed this in a 2008 publication (see Roberts *et al.*).

In time, other forthright scientists, also disturbed at the lack of fossil evidence for gradual evolution, began rethinking Drs. Schindewolf and Goldschmidt's hypothe-

ses. The first to do this publicly was Dr. George Gaylord Simpson in 1944 and 1949, renaming the rapid evolutionary change "quantum evolution." He believed that for evolution to occur it must have taken place in "sudden leaps" rather than gradually, so as not to be recorded in the fossils. Workers in the Soviet Union in the 1960s were also puzzled as to why over 100 years after publication of Darwin's *On The Origin of Species* no serious transitional fossils or missing links had yet been found. They also proposed that evolution must have occurred so rapidly that it left no geological evidence. It wasn't until thirty years after the "hopeful monster" story was penned that the idea gained widespread notoriety. It all stemmed from a 1972 publication by Dr. Steven J. Gould, Harvard Professor and Dr. N. Eldredge, American Museum of Natural History paleontologist, entitled, *"Punctuated Equilibria: An Alternative to Phyletic Gradualism."* They renamed the process, this time calling it "punctuated equilibrium" (which we will simply refer to as PE). The equilibrium was described as the vast eons of time in which no evolution took place, also-called stasis. This stasis was then punctuated (interrupted) by short episodes of wild, random, and beneficial mutations, creating new animal "kinds." Dr. Gould, in an article entitled "The Return of Hopeful Monsters," even prophesied, "I do, however, predict that during the next decade Goldschmidt will be largely vindicated in the world of evolutionary biology... The fossil record with its abrupt transitions offers no support for gradual change" (1977, pp. 22-30). **Translation:** "Traditional Darwinism is dead. We must come up with a new idea."

Dr. Eldredge accepted the PE position after his detailed studies of an ancient extinct arthropod, the trilobite. This creature's complex multifaceted eye, which some believe is far more advanced than the human eye, showed

SPECIATION

5

PUNCTUATED EQUILIBRIA: AN ALTERNATIVE TO PHYLETIC GRADUALISM

Niles Eldredge · Stephen Jay Gould

The first page of Gould and Eldredge's famous 1972 publication, first introducing the idea of punctuated equilibrium.

no signs of change over supposed millions of years of evolution, nor were there any serious transitional forms of the animal's body plan. This led Dr. Eldredge to conclude that evolution could not be proved by the fossils. In their landmark paper, Gould and Eldredge began by questioning the rationale behind evolutionists' refusal to deal candidly with the absence of transitional forms. Other well-known evolutionists soon joined the cast of PE hypothesis supporters including Dr. Steven M. Stanley, E.S. Vrba, and Derek V. Ager. Dr. Ager (in 1981), described the process in the following manner: "In other words, the history of any one part of the earth, like the life of a soldier, consists of long periods of boredom and short periods of terror" (pp. 106-107).

IV. The Reason Why Many Scientists Were Slow to Accept Punctuated Equilibrium

Many unwitting evolutionists accepted the new PE premise immediately. Many more rejected the idea, often because of the vulnerable position in which it placed them—they would have to admit that for the last 140 years they have been wrong. One scientist, weighing in on this issue, declared, "If evolution-related materials are changed to include the punctuated equilibrium

Dr. Gary Parker's concise response to macroevolution in which he shows that evolution couldn't happen fast, and it couldn't happen slowly.

concept, it amounts to nothing less than a tacit admission that creationists have been correct all along in stating that there is no fossil evidence" (Thompson, no date, *Issues in Evolution,* p. 16). Consider this. How quickly are evolutionary scientists, professors, and authors going to admit that the thousands of books, high school and college courses, scientific articles, lectures, and television programs over the past century and a half have been hoaxes? In spite of the facts, most evolutionists have come to accept the plausibility of PE in conjunction *with* conventional neo-Darwinian evolution—two ideas that

mix about as well as oil and water. Their reasoning goes something like this, "If the fossil record shows gradual changes and transitional forms, then traditional neo-Darwinian evolution must have occurred. On the other hand, if the missing links are not found, then this is clear proof that evolution took place by PE." What? As stated earlier, these are folks playing with a two-headed coin—*"heads I win, tails you lose."* Dr. Gary Parker summed up the creationist's view of this paradox most eloquently in explaining, "The 'rear guard' neo-Darwinian evolutionists like to point out the apparent absurdity of hopeful monster evolution and claim that evolution could not happen fast. The punctuational evolutionists point to genetic limits and the fossil evidence *to* show that evolution did not happen slowly. The creationist simply agrees with both sides: evolution couldn't happen fast and it didn't happen slowly" (Parker, 1994).

Twenty-one years after PE was proposed, Drs. Gould and Eldredge published their 1993 article, "Punctuated Equilibrium Comes of Age." In it, they cite examples of how PE has gradually been endorsed over the years stating, "punctuated equilibrium has been accepted by most of our colleagues… as a valuable addition to evolutionary theory" (pp. 223-227). And it has. On November 20, 2003 I attended a debate on evolution between Dr. Terry Mortenson (of *Answers in Genesis*) and Dr. Greg Hampikian, professor at Clayton State Univ. in Jonesboro, GA. In this debate, Dr. Mortenson made a compelling case for the lack of transitional fossils. Care to guess how Dr. Hampikian responded? He said that if the transitional fossils can't be found, then evolution must have taken place by PE—so fast that it left no trace in the fossil record. A few years back, I also corresponded with Dr. Donald Forsdyke, Professor Emeritus in biochemistry at Queen's University in Kingston, Ontario, Canada, a strong proponent of macroevolution. Dr. Forsdyke, commenting on the current view of Dr. R. B. Goldschmidt's ideas that led to PE said, "Goldschmidt is becoming increasingly seen as on the right track regarding his evolutionary ideas" (Forsdyke, 2005).

According to Gould and Eldredge, another benefit of the PE hypothesis is that, "palaeontologists never wrote papers on the absence of change in lineages before punctuated equilibri-

REVIEW ARTICLE

Punctuated equilibrium comes of age

Stephen Jay Gould & Niles Eldredge

The first page of Gould and Eldredge's 1993 article in Nature, vindicating punctuated equilibrium as a concept now accepted by evolutionists in addition to Neo-Darwinism.

um granted the subject some theoretical space" (1993, pp. 223- 224). This is true. The absence of missing links were once "hot potatoes," almost entirely ignored by Darwinists. This changed when the PE idea made the lack of transitional fossils "scientifically acceptable" to discuss. The absence of transitional forms was akin to an elephant in the living room that everyone knew was there but no one was willing to acknowledge, until PE was proposed. Their article concluded by stating "Darwinian extrapolation cannot fully explain large-scale change in the history of life" (*Ibid.*, p. 224). **Translation:** "Traditional evolutionary hypothesis is dead."

IV. Problems with Punctuated Equilibrium: No Fairy Godmother

So, where is the proof? All PE advocates readily admit that although no evidence can be cited to support their hypothesis, the process of rapid evolutionary change had to have occurred because it is the only alternative to a lack of transitional fossils. Gould himself admitted that the process of PE sounded more than a little absurd, and yet admitted that we have to believe it because it "must have occurred." He asked, "How could we ever convert a rhinoceros or a mosquito into something fundamentally different? Yet transitions between major groups must have occurred in the history of life" (1977, pp. 22-30).

A. The time problem of Punctuated Equilibrium. The proposed time period necessary for PE to occur has never been agreed upon. The suggested time, claimed necessary for PE

to have taken place, varies from one year to ten million years, depending on who is speaking, which is only a blink of an eye in evolutionary time, which supposedly requires hundreds of millions of years. The problem with time in PE is twofold. First, to assume that a species mutated into another living kind rapidly is, by the evolutionists, admissions, illogical and statistically impossible. However, to accept more protracted periods of tens of millions of years creates this conundrum: Why do transitional fossils not appear in the geologic record? Truly, the process could not have occurred in any amount of time.

B. The absence problem of Punctuated Equilibrium. PE enthusiasts are quite right to point out the absence of transitional fossils in the geological record. What they are not so quick to admit is the fact that there is also an absolute absence of evidence supporting PE in all of science. It simply is not there. And this is where things get very strange. The fact that there is no evidence in the fossil record is said to be *proof* that PE occurred. "It occurred so *quickly* that it didn't leave any evidence" they tell us. Really?? This marks the first time in all of science that a widely accepted theory has been formulated around and accepted based on an entire lack of any supporting facts. Here's an example. Dr. Derek Ager said, "Gould constantly emphasized that 'stasis is data,' in other words nothing happening in a fossil lineage is just as important as something happening" (1993, p 131).

C. Genetics: the mechanistic problem of Punctuated Equilibrium. The occurrence of an enormous mutational punctuation in the established genetic succession of a group of life forms is entirely undocumented, unprecedented, and unscientific. PE defies everything ever learned about how genes are passed on to successive generations. Dr. Alan Hayward said, "Genes seem to be built so as to allow changes to occur within certain narrow limits, and to prevent those limits from being crossed" (Hayward, 1985, p. 19). Speaking to this problem, evolutionist Dr. Sewell Wright recollected, "I have recorded more than 100,000 newborn guinea pigs and have seen many hundreds of monsters of diverse sorts, but none were remotely 'hopeful,' all having died shortly after birth if not earlier" (1982, p. 440). Dr. Wright is reminding scientists that an animal born with

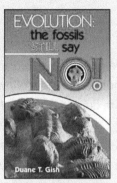

Former Cornell professor, Dr. Duane Gish has been called "creation's bulldog" for his tireless defense of biblical creation and the young earth. Dr. Gish had over 350 debates with evolutionists, was the past president of the Institute for Creation Research, and passed away in 2013. "Creation's bulldog" is a play on T.H. Huxley's nickname, "Darwin's Bulldog."

Dr. Duane Gish's classic book that exposes the flaws of macroevolution from the fossil record.

multiplied mutations would not survive long past birth, much less be able to procreate. Former Cornell University biochemist Dr. Duane Gish illuminated this fact in a lengthy quote, we have shortened for the sake of space: "The genetic apparatus of a lizard... is devoted 100% to producing another lizard. The idea that this... amazingly stable genetic apparatus... could be drastically altered... in such a way that the new organism not only survives but actually is an improvement over the preceding form is contrary to what we know" (1995, p. 335).

Conclusion

Is punctuated equilibrium evolution's new hope? This is simply not possible, and anyone who teaches otherwise is obviously living in a world where dreams become possibilities, possibilities become probabilities, and probabilities become proven science. This is the sad world inhabited by those who deny the existence of God and the facts of his miraculous creation (Ps. 14:1). In a discussion of this fairytale alternative to traditional evolution, Dr. Henry Morris said, "It would seem that one could as easily believe in a fairy godmother with a magic wand!" (Morris and Morris, 1996, p. 105).

NOTES

Questions
FILL IN THE BLANK

1. Approximately _____ species of fossils have been unearthed with no missing links found.
2. Dr. _____ proposed the "hopeful monster" idea, following in the footsteps of his predecessor Dr. _____, both of whom denied traditional Darwinian evolution.
3. Based on years of studying the eyes of _____ fossils, Dr. Niles Eldredge rejected gradual neo-Darwinian evolution and proposed the new concept of punctuated equilibrium.
4. What did George Gaylord Simpson rename the "hopeful monster" view?_____
5. Evolutionists say that the _____ of transitional fossils actually proves _____ _____.
6. "The fool has said in his heart _____ _____ _____ _____" (Ps. 14:1).

Short Answer
1. What passages teach us that sin can blind us? _____
_____

2. What passages teach us that sin can harden us?_____

3. Explain the hopeful monster/punctuated equilibrium idea. _____
_____
_____

4. What might be some of the reasons that many evolutionists shy away from wholly accepting PE and wholly rejecting the time-honored neo-Darwinian mechanism of evolution? _____
_____
_____

5. Explain why PE could not have occurred in either a million years, or tens of millions of years. _
_____
_____

6. Why do you think that many evolutionists, such as Dr. Gould, conclude that PE *"must have occurred"*? _____
_____

Yes or No
____ 1. Saltation evolution (saltationism) teaches that mutations gradually accumulated in animals over vast amounts of time.
____ 2. Since the nineteenth century, scientists have expressed their disagreement with Darwinian evolution.

_____ 3. Intelligent and scholarly individuals are never misled.

_____ 4. Atheistic evolutionists have much incentive to be straightforward and honest with the facts.

_____ 5. There is no difference between the old "hopeful monster" view and the, now accepted, idea of punctuated equilibrium.

_____ 6. Many scientists today accept the possibility of PE as an evolutionary mechanism.

Discussion Question in Preparation for Answering Unbelievers and Critics

While on a field trip to a cave, the park ranger tells the class that many fossils have been found there. She says, "Although no undisputed fossilized missing links have been found, there is other evidence that shows how evolution may have occurred. For instance, evolution could have occurred so rapidly by punctuated equilibrium that almost no transitional fossils were left in the earth." She then asks if there are any questions. What do you say? _____

_____

_____

_____

_____

_____

_____

Lesson 11

Catastrophism, The Global Flood, and the Age of the Earth (Uniformitarianism Disputed)

Introduction

Neo-Darwinian evolution avows that millions of years are necessary to accomplish its purpose. Thus, a study of the age of the earth is entirely in order. Although the biblical record suggests that the earth is "young," only a few thousand years old (see Lessons four and five), evolutionary proponents, even some members of the church, argue for an "ancient" earth (billions of years old) in order to be in step with the evolutionary scientific community. However, if it could be proven that the earth is relatively young, a blow to the evolutionary theory would be delivered so as to destroy the very bedrock of Darwinism's creative force... *time*. For this reason, evolution *cannot* allow for a young earth.

"The Deluge," oil painting in the Tate Gallery, London. By Francis Danby (1793-1861). Painted circa 1837-1839.

Left to right: James Hutton (1726-1797) and Charles Lyell (1797-1875) who made famous the teaching of uniformitarianism and heavily influenced a young Charles Darwin.

I. Uniformitarianism versus Catastrophism

The doctrine of uniformitarianism maintains that there has been uniformity in all natural, physical, and geological events ever since the inception of the earth. That is, the world we see around us is all a consequence of unhurried change by the forces of nature over billions of years. The teaching says that just as natural events gradually transpire today, natural events have always taken place in the same slow and gradual fashion in the past. Two key British and Scottish scientists, Hutton and Lyell, beginning with the premise that the Bible is inaccurate and does not represent a true record of the earth's history, made popular the uniformitarian doctrine (MacArthur, 2001, p. 44). The teaching of these two men heavily influenced the young Charles Darwin who brought a copy of Lyell's uniformitarian book, "The Principles of Geology", with him on the HMS Beagle in 1831 (*Ibid.*, p. 51). In his book, Lyell made popular, to the common man, the idea that the geologic column was laid down in a uniformitarian fashion, very slowly and gradually over millions of years. This stood in stark contrast to the prevailing idea of the day, which was that all the sedimentary layers of rock were laid down rapidly based on "catastrophism." Uniformitarianism not only denies that supernatural events were involved in creating and shaping the earth, but it also rejects the teaching that great catastrophes such as a global flood, great earthquakes, and volcanic eruptions helped carve the earth's texture and create the geologic column. The apostle Peter censures such anti-religious philosophy as uniformitarianism in saying, "Knowing this first: that scoffers will come in the last days, walking according to their own lusts, and saying, 'Where is the promise of His coming? For since the fathers fell asleep, all things continue as they were from the beginning of creation.' For this they willfully forget: that by

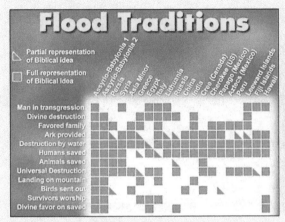

Every major culture around the world has a flood tradition or legend, many very closely matching the events that are recorded in Genesis 6-8. Other cultures with detailed flood traditions include Australia, Bolivia, Borneo, Burma, Canada, Cuba, East Africa, Egypt, French Polynesian, Guyana, Iceland, Iran, Malay Peninsula, Mexico (Codex Chimalpopoca), Mexico (Huichol), Vanuala, Vietnam, Wales, Alasaka (Tlingit), Alaska (Kolusches), New Zealand, etc. (Diagram: Courtesy of Northwest Creation Network).

An artist's depiction of the Noahic flood that may have led to the burial of dinosaur carcasses under rapidly-layered sedimentation. (Illustration by Lewis Lavoie via Apologetics Press. Used with permission).

In Noah's Ark: A Feasibility Study, Dr. John Woodmorappe presents highly technical arguments for the ability of Noah's Ark to carry out the mission that God assigned it. Chapters deal with inner dimensions of the ark, how many animals were on the ark, waste management, feeding and watering animals, end of flood events, spread of animals after the flood, etc. (Woodmorappe, 2009).

the word of God the heavens were of old, and the earth standing out of water and in the water, by which the world that then existed perished, being flooded with water" (2 Pet. 3:3-6).

The Bible suggests that a quite different scenario than uniformatarianism is responsible for the hydrological and geological features on the earth; namely, catastrophe. The Bible describes at least two major worldwide upheavals that account for the greatest sculpting of mountain, land, and sea. The first event occurred during the first two days of creation (Gen. 1:1-10). Moses recounts, "Then God said, 'let there be a firmament in the midst of the waters, and let it divide the waters from the waters.' Thus God made the firmament, and divided the waters which were under the firmament from the waters which were above the firmament; and it was so.... Then God said, 'Let the waters under the heavens be gathered together into one place, and let the dry land appear;' and it was so" (Gen. 1:6-9). The second major event that occurred was the global flood, when it not only rained for forty days and nights, but "all the fountains of the great deep were broken up" (Gen. 7:11). And in Genesis 7:19 we read, "The water prevailed more and more upon the earth, so that all the high mountains everywhere under the heavens were covered" (NASB). These two great events would have probably involved massive tectonic plate shifts, volcanic activity, including volcano-induced steam geysers, and ensuing millions of tons of sedimentary depo-

sition. The consequences would be the rapid layering of sediment that is seen in the geologic column, along with the millions of fossils, and vast stores of coal from the compression of many cubic miles worth of organic materials, such as animals and vegetation.

We should make a comment here about mass extinctions. Evolutionists claim there were at least seven separate mass extinctions of animals that took place, as seen in the fossil record. Creationists claim there was one great flood that buried billions of animals in different layers, now interpreted by evolutionists as separate extinctions. Evolutionist Dr. Derek Ager said, "Probably more significant and difficult to explain, was the complete collapse at this time of the whole marine ecosystem from the minute coccoliths via the ammonites and belemnites to the large marine reptiles" (Ager, 1993b, p. 181).

Another question that frequently arises is, "How did Noah get all those animals in the ark?" The answer is fairly simple, if you consider just a couple of facts. (1) The inner dimensions of the ark were approximately 467,000 cubic feet of space. This is equivalent to 173 freight carriers on each of 3 decks of the ark = a total volume of 1.4 million cubic feet of space! This is equivalent to 500 freight carriers or 290 semi trailers. As to the animals, baraminologists (creation scientists who try to determine what the original "created kinds" of animals were), conclude that there are only ca. 2,000 - 8,000 different "created kinds" of animals (Woodmorappe, 2009). This means only 4,000 - 16,000 animals had to be taken on the ark. For large animals, juveniles could be taken. For the dinosaurs, all species can be distilled into only fifty different original "created kinds" of dinosaurs. Then, the dinosaurs could either have been brought on board as juveniles or as eggs, which were no larger than the size of a football. In fact, all of these animals could fit on only one deck of the ark, leaving two more decks for storage, food, waste, and dwelling areas for the eight humans.

Catastrophic occurrences could account for geologic findings such as the "Karoo Supergroup" of Africa that contains the fossils of around 800 *billion* rapidly buried vertebrate animals. It can also easily account for the Miocene shales of California where within a four square mile area, one *billion* fish were fossilized. Uniformitarianism cannot account for these events. Nor can it account for the creation of the Columbian Plateau lava beds in this country that cover 200,000 square miles and are thousands of feet thick (Gish, 1995, pp. 48, 49). Catastrophism easily explains all of these phenomena. Further, consider that until about the year AD 1800, all the geological formations on the earth were interpreted in light of the global flood of Noah and catastrophic activity in renowned universities such as Yale, Harvard, Oxford, and Cambridge (*Ibid.,* p. 49). Only in the nineteenth century's mad rush to discredit the Bible did uniformitarianism take root.

One more note on the geologic column. If there really were a great Noahic flood, it would only stand to reason that an organized succession of fossil evidence would be found, based on the abilities of animals to escape the ascending tide. Creation scientists predict that the progressive layering of the remains of life forms from the bottom up might occur in this order: (1) small and immobile oceanic creatures → (2) larger and swimming sea creatures → (3) seashore and river animals → (4) smaller mammals, reptiles, amphibians and lowland forests → and finally (5) large mammals, birds, animals of higher intelligence and upland forests. This is precisely the order we see in the geologic column. However, one might ask, "Then where are all the human fossils?" Good question. If we accept the biblical account as true, it is feasible for ~ten million people to have been living on the earth at the time

The Karoo Valley of Desolation, in South Africa, located amidst the Karoo Supergroup containing around 800 billion rapidly buried fossils. (Photograph by Dinal Cheal. Used with permission).

of the flood in Genesis chapter seven. If about half of these bodies were completely preserved and evenly distributed throughout the ~700 million cubic kilometers of fossil-bearing rock, one human fossil would be found in only every 140 cubic kilometers of rock (Batten, 2003, pp. 195, 196). Hence, the probability of finding even one human fossil from the flood is incredibly small.

Geologic folds in the Miocene shales of California where within four square miles a billion fish were rapidly buried. (Photo by Dr. Allen Glazner. Used with permission).

II. "But It Looks So Old!"

So say many young earth critics. "If the earth is really only a few thousand years old, then why does it *look* like it is billions of years old," they ask. Consider this. According to Genesis one, when God created the earth and all that is in it, did He create a mature earth and mature creatures, or infantile ones? We are told that when God made man and woman He told them to "be fruitful and multiply" (Gen. 1:28). But, how old was Adam one hour after God created him? One hour old. How old must he have ap-

peared? He would have appeared older than one hour because God created him a grown man of reproductive age. Animals were commanded to be fruitful and multiply too—obviously they were created mature enough to reproduce (Gen. 1:20-22). What about the trees that provided fruit in the garden? The trees were fruit-full, appearing to be years old immediately after creation. A full grown elephant, a day after creation, might have had an apparent age of fifty+ years, although it was only twenty-four-hours old. Thus, it is unreasonable for one to insist that the earth must *appear* young in order for it to *be* young. God created the world in a mature state. No trickery or deception was involved. If someone assumes the earth is old because it *appears* old to them, then he must discredit the first two chapters of the Bible. This is precisely the goal of the humanist, the Darwininan evolutionist, and the modernist theologian. Notice this admission from an atheist, "Christianity has fought, still fights, and will continue to fight science to the desperate end over evolution, because evolution destroys utterly and finally the very reason Jesus' earthly life was supposedly made necessary. Destroy Adam and Eve and the original sin, and in the rubble you will find the sorry remains of the Son of God. If Jesus was not the redeemer who died

for our sins, and this is what evolution means, then Christianity is nothing" (Bozarth, 1979).

The question should also be asked, by what standard do you interpret an "old looking earth" from a young one? If the earth was created in a great catastrophic event, followed by a global flood, followed by ~4,000 years of weathering and deterioration, would it not appear as we see it today? Critics counter that the earth was not created miraculously, or if it was, it was created slowly by a molten sphere of lava cooling and then undergoing a total of 4.54 billion years of evolution. How can one know this? By who's standard does this earth "appear" old? Is it man's place to play God and tell Him how the earth began? The patriarch Job did just this, questioning and challenging God. God responded, in short, by asking, "Where were you when I laid the foundations of the earth?" (see Job 38). Many so-called scientific experts of the past, along with their ideas, have long been discredited and supplanted by successive generations of scoffers devising newer theories that will only later be disproved by others. Apparently, they didn't learn the lesson that Job did… "Were you there?"

III. Evidence for a Young Earth and Cataclysmic Sedimentation

Although God's word should be sufficient in this regard, we will provide a few examples of the great weight of evidence in favor of a young earth.

A. Contemporary rapid sedimentation. Uniformitarianism teaches that millions of years are necessary to produce the sedimentation we see in the geologic column. Not so. Mount St. Helens erupted in Washington State in 1980 producing finely layered sediment 25 feet deep in a single afternoon (MacArthur, 2001, p. 53; Sarfati, 1999, p. 105). If it is possible that 25 feet of sediment could be formed in a few *hours*, imagine what a global flood would do in a period of *months.* Dr. Guy Berthault, as well as other evolutionist researchers, conducted multiple experiments modeling what would happen in a worldwide hydrological catastrophe. They reported that when various sizes of sediment are mixed in water and allowed to settle, a self-sorting phenomenon occurs in which separate layers of rock are formed in the exact same pattern regardless of the flow rate of wa-

Interbeded Cambrian and Mississippian layers of sediment dispel the uniformitarian evolutionist's perceived construction of the geologic column. (Photo courtesy of Don Patton and Steve Rudd. www.bible.ca)..

ter. I have an excellent video demonstrating this phenomenon. Thus, a worldwide flood would produce similar layers of sedimentary rock in all parts of the earth (Safarti, 1999, p. 106). One problem for the evolutionists is that some sedimentary layers, that were supposedly created millions of years apart, are often *interbedded* within one another. One example of this shows Cambrian and Mississippian sedimentary layers interbedded one on top of the other. The problem with this is that the Cambrian (Mauv limestone) layers were allegedly laid down over supposedly 150 million years before the Mississippian (Redwall limestone) layers! How did the Cambrian come out on top?

B. Evidence of rapid sedimentation. Evidence exists that the sedimentary layers in the geologic column were formed rapidly, rather than slowly. This can be seen in the deformation of soft pre-solidified rock layers with things such as fossilized water ripples, animal tracks, and even rain drops. Cliffs and mountain sides often show the bending, folding, and reforming of sedimentary rock, which evolutionists claim occurred over millions of years due to tectonic plate disruption. If these layers were formed by slow upheavals of rock masses, then these formations would not appear as we find them today—without any evidence of cracking, heating, or melting. This suggests that the deformation of layers took place while the layers were still soft. There is also scant evidence of any erosion, soil formation, animal burrows, or roots

between layers that should be abundant if sediment was slowly deposited over eons of time (Batten, 2000, p. 156). Other evidence of rapid rock formation includes the formation of clastic dykes and pipes or cylinders where a sand and water mixture has forced its way up through multiple sedimentary layers of sandstone while the sediment was still soft (*Ibid.*, pp. 156, 193). Finally, rocks holding dinosaur fossils contain only the dinosaurs and not the vegetation necessary to sustain life. But, why? (*Ibid.*, p. 194). Where is all of their food?! A full grown *Apatosaurus* would require over three tons of vegetation per day! Why are the dinosaurs buried absent of their surrounding ecosystem? Could this be evidence of a catastrophic burial? Yes. In March of 2008, a fantastic discovery of the largest dinosaur burial pit ever found was discovered in Zhucheng, China. Apparently, 7,000 dinosaurs were all washed together and rapidly buried in sediment in order to preserve their bones before decomposition could take place. It is estimated that there are up to 15,000 dinosaur bones in one 980 foot ravine. Although this fits in perfectly with the Noahic flood model, evolutionists aren't completely agreed on how this happened. Today up to 1,500 dinosaur bones can be seen in one exhibit still buried in the rock where they died in the Chinese ravine. Again, this is a rapid burial of diverse dinosaurs, in addition to turtles, crocodiles, and clams. Similarly, Dinosaur National Monument was first discovered in 1909 on the border of Colorado and Utah, with an 80 acre burial pit where dinosaurs were rapidly buried in a watery slurry of sediment. Up to half of all dinosaurs thought to have lived in North America are buried together there.

C. Polystrate fossils. Polystrate fossils are remains that traverse multiple layers of sedimentary rock. Remember, these are layers that were supposedly laid down over millions of years. The millions of polystrate fossils that have been uncovered prove that life forms were trapped in

A polystrate fossilized tree near Cookville, TN in the Kettles coal mines. This tree was supposedly buried gradually amidst millions of years of sediment. Only a rapid catastrophic burial could create such pristine preservation. (Photograph courtesy of Don Patton and Steve Rudd. www.bible.ca).

A polystrate lycopsid, in Cumberland Basin, Nova Scotia. Lycopsids were large tree-like plants from the age of the dinosaurs, which they may have used for food. This lycopsid, was buried rapidly in a catastrophic flood, and was petrified. Notice the details of the roots at the bottom (Photo, Rygel, see references).

A paleontologist at Dinosaur National Monument, excavating dinosaur vertebrate so visitors can see the fossils in the rock, where the animals died in a great catastrophic flood (Photo: USGS).

The "Fossil Tunnel" in Zhucheng China, where visitors can see some of the 7,000 specimens of dinosaurs rapidly buried in a flood and preserved in the walls of the ravine. (Photo credit: Glensmart, in references).

sediment that formed around them rapidly, rather than slowly. Dr. Carl Wieland, in his book *Stones and Bones* (available online), published photographs of many fossils that could only be the product of rapid sedimentation that would have had to occur prior to any significant decay of organic tissues, which begins soon after death. These photographs include those of a seven-foot long ichthyosaur buried in multiple layers of rock while giving birth, a fish buried in sediment while feasting on a smaller fish, a jellyfish fossil, and an upright tree, similar to those which have been discovered throughout the world, traversing supposedly millions of years of sedimentary rock (Wieland, 1994). Is it plausible to assume that these polystrate trees remained standing undecayed for millions of years while sediment incrementally gathered around them? Or, is it more reasonable to conclude that these giants were rapidly buried in a worldwide cataclysm? Remember that wood, made of organic compounds, could not survive for millions of years and begins decaying soon after death. Finally, the most famous polystrate fossil is that of a baleen whale that is more than seventy-five-feet long. It stands almost on its tail at a 60° angle through supposed millions of years of sedimentary layers and was reported in a major scientific journal (Reese, 1976, p. 40). Hear from the Professor and Head of the Department of Geology, University of Swansea, evolutionist and agnostic, Dr. Derek Ager, "Probably the most convincing proof of the local rapidity of terrestrial sedimentation is provided by the presence in the coal measures of trees still in position of life. Two Late Carboniferous trees stand in the garden of Swansea Museum... Such stand trees are not uncommon in the Upper Carboniferous" (Ager 1993b, p. 47). "Broahurst and Loring recorded standing trees up to ten meters high in the Lancashire coalfield of northwest England... Obviously sedimentation had to be very rapid to bury a tree in a standing position before it rotted and fell down... Standing trees are known at many levels and in many parts of the world. Thus there are trees... which were evidently buried by a sudden rush of sediment. In the English Upper Jurassic we have the 'fossil forest'... where cycad-like trees stand in an ancient soil... At Yellowstone National Park in Wyoming, a whole forest of Miocene trees is buried in volcanic ash. Similarly in the Ginkgo Petrified Forest near Vantage in Washington State, many different Miocene trees are still in position, preserved between lava flows" (*Ibid.*, p. 49-50).

D. Out-of-place fossils. Many fossils found in incongruous locations illustrate that man coexisted with animals that we are told disappeared millions of years prior to man. William J. Meister, in 1968, uncovered the footprint of a sandal-wearing human with a supposedly half a billion year old trilobite pressed right into the track (Cook, 1970, pp. 186-193). A metal hammer with a wooden handle was excavated out of supposedly 135 million year old Cretaceous limestone (Helfinstine and Roth, 1994). Other out-of-place fossils include a fossilized leather sole imprint with sown stitches in Triassic rock that is supposedly 225 million years old, a fossil sole imprint with sewed thread in coal supposedly fifteen million years old, and flint carvings on extinct reptile (saurian) bone, supposedly 180 million years old (Von Fange, 1974, p. 19).

E. Red blood cells and fleshy tissue in dinosaur bones. Unfossilized dinosaur bones containing red blood cell components have been discovered (Schweitzer *et al.*, 2005, 2005a, 2005b, 2009, Stokstad, 2005). Others have discovered a dinosaur egg with a soft leathery shell and an embryonic skeleton with soft tissue attached inside [Ji *et al.*, 2004], undigested soft muscle tissues within the feces of a large Cretaceous tyrannosaurid [Chin *et al.*, 2003], and detailed soft tissue preserved from animals of the early Cretaceous Jehol Group (Zhou *et al.*, 2003). The ages of these remains are a few thousand years at best and certainly not the scores of millions of years old that evolution avows.

F. Thick tightly bent rock strata, with no melting, fracturing or sand grain elongation indicates that the rock was bent while it was still soft, wet and pliable. Examples of geological layers bent while they were still soft, wet, and pliable are found around the world. One example is the Kaibab Upwarp in the Grand Canyon. Basement layers of rock there indicate melting, fracturing, and re-crystallization. However, the higher layers indicate they were bent before they hardened (Batten, 2009; Walker, 2015). Creation scientists believe this occurred following Noah's flood.

G. Carbon-14 found in diamonds, coal, and fossils. The isotope Carbon-14 (C-14) gives off subatomic particles so that it has a half-life of 5,730 years (i.e., it loses 50% of its mass every 5,730 years). Thus, it should be undetectable after 40,000 years (Bowman, 1990) and completely gone after 1 million years. Nevertheless, C-14 has been found in diamonds that are supposedly 1-3 billion years old, and fossils and coal that are a supposed 32-350 million years old (Snelling, 2012). (NOTE: This has even been confirmed by secular scientists. See Appendix F for

Photograph and close-up photograph of supposed 500 million year old trilobite pressed into the heel of a human footprint. (Photos by Brad Harrub via Apologetics Press. Used with permission).

ninety ancient geological samples that are supposedly millions of years old and thus should contain no more C-14. However, they are now shown to contain C-14 and be less than 40,000 years old by C-14 dating).

H. The Distance of the Moon from the Earth. The moon is receding from the earth at a rate of 1½ inches per year. However, if the earth is on the order of five billion years old, then the moon should be at least 3½ times further from the earth than it is presently (DeYoung, 1990, pp. 79-84; Batten, 2009).

I. The salinity of the oceans. The salinity in our oceans gradually increases over time due to salt, rapidly pouring in from mineral runoffs in rivers and streams. Every year, 458 million tons of salt are added to the oceans (Meybeck, 1979), while only 122 million tons are removed (Sayles and Mangelsdorf, 1979). However, if the earth is as old as evolutionists claim, then the salinity of our oceans should be eighty times greater than the 3.5% they currently are (30; Batten, 2009). Evolutionist Dr. C.B. Gregor said, "If magma

Fossilized hammer discovered in supposedly 135 million year old rock near London, TX. (Photo courtesy of Don Patton and Steve Rudd. www.bible.ca).

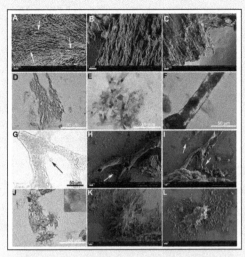
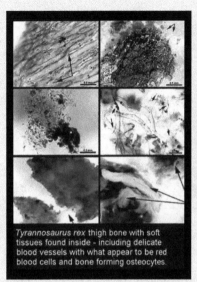

Micrographs of supposed 80 million year old "Duck Bill Dinosaur" (Hadrosaur) soft tissue found by Mary Schweitzer (2009). The pictures are (A) Dinosaur bare cells called osteocytes, (B) Bone matrix fibers, (C) Ostrich bone fibers for comparison, (D) Dinosaur bone matrix, (E) Dinosaur blood vessels and red round inclusions, (F) Zoomed in photo of blood vessel with red inclusions, (G) Ostrich blood vessel for comparison, (H) Hollow dinosaur blood vessels, (I) Ostrich vessel for comparison, (J) and (K) Osteocytes with protein fibers called filopodia, (K) Ostrich osteocyte for comparison. NOTE: The authors detected eight peptides with 149 amino-acids from four different samples.

Tyrannosaurus rex thigh bone with soft tissues found inside - including delicate blood vessels with what appear to be red blood cells and bone forming osteocytes.

Photos: Schweitzer et al., 2005a, with editing by Dr. Sean Pitman accessed on 4/11/15 at: www.detectingdesign.com

Dr. Mary Schweitzer who discovered soft tissue in T. Rex bone and has published over 30 peer-reviewed scientific publications concerning non-decomposed tissue in dinosaur bones. (Photo Credit: nwcreation.net).

kept the crust built up against the ravages of erosion and the waste products accumulated in the sea, at present rates of influx the ocean basins should long ago have been choked with sediment and salt ... . salt must somehow leave the ocean" (1988). Further, assuming a starting percent of zero and maximizing all the rates of input, the oceans would only be sixty-two million years old, less than 1/60 their supposed age of 3.8 billion years old, claimed by evolutionists. For a mathematical model of ocean salt exchange, see Austin and Humphreys (1990).

J. Ocean sediment deposition. Every year, twenty million tons of sediment are washed into the oceans, while only one million tons are removed by tectonic plate subduction. For a mathematical model of this process, see Vardiman (1996). Were the oceans actually 3.8 billion years old, sediment buildup would be exceedingly great. Some estimate, however, that an average of only approximately 3,000 feet of sediment has accumulated from these processes giving the earth a maximum age of only ~10,000 years (Williams, 1996, p. 62, Batten, 2009). In fact, much of the world's sea floor has no sediment at all, as assessed by the NOAA National Geophysical Data Center sediment thickness map (Divins, 2003). Consider the mighty Mississippi River that deposits

~300 million cubic yards of sediment into the sea per year (Thompson and Harrub, 2003, p. 6). If the earth were as old as evolutionists' claim, the Gulf of Mexico would have long ago been filled in with debris.

K. The Grand Canyon. Evolutionists claim that the Grand Canyon was created through the erosive force of the Colorado River over millions of years. One of the many problems with this hypothesis is that there is over 5.45 trillion cubic yards of sediment missing (*Ibid.,* p. 6, National Park Service, 2015). Where did it go? Only catastrophic processes can adequately explain the creation of this wonder of the world.

L. Decay of the earth's magnetic field. The earth's magnetic field is naturally and rapidly disintegrating (Batten, 2009). In fact, one government report estimated that the end of the earth's magnetic field will occur around the year AD 4000. Calculating backwards to the maximum possible electrical force, Dr. T. G. Barnes of the University of Texas estimated the maximum age of the earth to be around 10,000 years (1981). In "Earth's Magnetic Field," Wikipedia claims that the Earth's magnetic field has declined 10-15% in the last 150 years and 35% in the last 2,000 years (which shows that the decay is speeding up). The article claims a decay rate of 6.3% per century and at this rate, it says the magnetic field would disappear in 1,600 years. However, the evolutionist's rescuing device argues that the magnetic field

Mountain layers bent while they were soft and pliable at Eastern Beach near Aukland, New Zealand (Photo Credit: Batten, 2009).

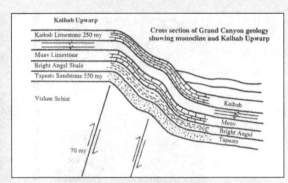

Kaibab Upwarp in the Grand Canyon where the Vishnu basement layers of rock there indicate melting, and fracturing while being bent and then re-crystallization after being bent. However, the higher layers (Tapeats Sandstone, Bright Angel Shale, Muav Limestone and Kaibab Limestone) indicate they were bent while they were soft, wet and pliable, before they hardened (Picture credit: Walker, 2015).

will inexplicably increase at some time in the unknown future. This has been disproved by detailed mathematical models published by Dr. Russell Humphreys (physicist, Sandia National Laboratories) (Humphreys, 1990, 1993, 1996).

M. Short-lived comets could not survive for billions of years in our solar system. Comets are light objects circulating in the solar system, composed of ice, dust, and frozen gases. As comets orbit the sun, each pass melts away a significant portion of the comet, leaving a long and often beautiful tail that can be seen from satellite telescopes and sometimes from the earth. Therefore, the maximum age of short-lived comets is estimated to be 100,000 years, thus they would not survive the supposed 4.6 billion year age of our solar system. This means that our solar system must be less than 100,000 years old. Not to be outdone, the evolutionists have invented a rescuing device, saying that there must be an "Oort Cloud" of comets surrounding our solar system and every now and then, the Oort Cloud kicks a new comet into the orbit of our solar system. Yet, there has never been a trace of evidence of the existence of the Oort cloud. Writing in the journal *Nature* in 1989,

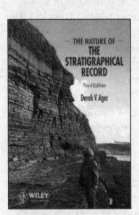

Dr. Derek Agar's book, The Nature of the Stratigraphical Record (1993a) establishing that the doctrine of uniformitarianism, taught since 1800, has been wrong and that the geologic column was formed by catastrophic processes.

John Maddox stated that Halley's Comet has a maximum possible age of 23,000 years and has only orbited our solar system ca. 300 times. Atheistic cosmologist Carl Sagan and his wife Ann Druyan said, "Many scientific papers are written each year about the Oort Cloud, its properties, its origin, its evolution. Yet there is not yet a shred of direct observational evidence for its existence" (Druyan and Sagan, 1997). The number of comets in our system is continually decreasing, demonstrating our solar system is thousands, not billions of years old.

N. The Presence of DNA and revival of supposedly 250 million year old "Lazarus Bacteria." Researchers isolated ancient bacteria from salt crystals dated to 250 mya in a cavern in Carlsbad, NM, which comprises a new species never seen before, which they call, *Bacillus permian*. Detractors say the bacteria must have been in the salt as a result of contamination by the researchers. However, publishing in the prestigious journal *Nature*, Vreeland *et al.* (2000) stated that sterilization procedures reduced the possibility of contamination to less than a 1 in 1 billion chance. The question is, how could bacteria live this long?

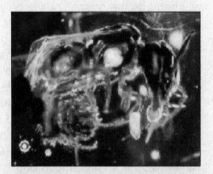

An extinct bee (dated by evolutionists at 25-40 million years old) trapped in amber from which bacteria was revived from the bee's digestive tract (Cano et al., 1995). It is more logical to believe that the bee was trapped in amber shortly after the Noahic flood, some 4,500 years ago.

250 million years? Really? It is more probable to assume that the bacteria was trapped in the salt crystals in the cavern after the Noahic flood, approximately 4,500 years ago. Even evolutionists admit that DNA in a bacterial spore should last less than 1 million years. (See also Coghlan, 2000; and Vreeland *et al.*, 2002). Further, bacterial spores were isolated from an extinct bee trapped in amber dated at 25-40 million years old (Cano *et al.*, 1995).

IV. Newsflash: Evolutionists Agree, the Geologic Column Was Formed by Catastrophism and Not Uniformitarianism

What? Could it be true? After 220 years of geological teaching that the geologic column was laid down gradually (inch by inch) could the evolutionists really be reversing their course? The fact is that since the 1960's and 1970's, scientists have begrudgingly admitted that the geologic column had to be laid down by catastrophic processes *rapidly* and not by incremental additions (inch by inch). One of the leaders in this field was Dr. Derek Agar (Professor and Head of the Department of Geology and Oceanography, University College of Swansea), who admitted, "To me, the whole record is catastrophic, not in the old-fashioned apocalyptic sense... but in the sense that only the episodic events—the occasional ones—are preserved for us" (1984, p. 93). Later he stated, "In the late Carboniferous Coal Measures of Lancashire, a fossil tree has been found, thirty-eight feet high and still standing in its living position. Sedimentation must therefore have been fast enough to bury the tree and solidify before the tree

had time to rot... The hurricane, the flood or the tsunami may do more in an hour or a day than the ordinary processes of nature have achieved in a thousand years. Given all the millennia we have to play with in the stratigraphical record, we can expect our periodic catastrophes to do all the work we want of them" (1993a, pp. 65, 68, 69). Other eminent scholars have agreed with Dr. Ager, accepting the view of "neo-catastrophism." See Appendix D for more quotes from eminent scientists admitting that the geological column was laid down in catastrophic rather than uniformitarian processes.

Conclusion

Time fails us to list more evidence in favor of a young earth. However, for the Christian, God's word should forever settle the controversy of the age of the earth. "For in six days the Lord made the heavens and the earth, and on the seventh day He rested" (Exod. 31:17).

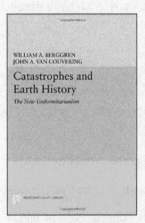

Berggren and Van Couvering's edited book Catastrophes and Earth History (2014), which admits that the doctrine of uniformitarianism, taught since 1800, has been wrong and that the geologic column was formed by catastrophic processes.

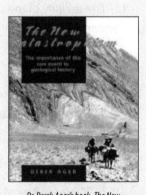

Dr. Derek Agar's book, The New Catastrophism (1993b) establishing that the doctrine of uniformitarianism, taught since 1800, has been wrong and that the geologic column was formed by catastrophic processes.

NOTES
_____
_____
_____
_____
_____
_____

Questions

MULTIPLE CHOICE

Choose all that apply

1. Examples of rapid sedimentation in the geologic column include: (a) Clastic dykes; (b) Deformation of soft pre-solidified rock layers; (c) Erosion between layers; (d) Cracking of layers during deformation.

2. Examples of polystrate fossils include: (a) An ichthyosaur buried amidst multiple layers of sediment while giving birth; (b) A fish buried in sediment while eating another fish; (c) Upright trees transversing supposed millions of years of any sedimentary layers; (d) A seventy-five-foot long whale buried through multiple layers of sediment.

3. Examples of out-of-place fossils include: (a) A trilobite pressed into a sandal imprint; (b) A hammer in Cretaceous limestone; (c) A human sole imprint in coal; (d) Flint carvings on a saurian bone.

4. Catastrophism can account for the following geologic findings: (a) The Karoo Supergroup of Africa; (b) The Miocene shales of California; (c) The Columbian Plateau lava beds; (d) The Suez Canal.

5. Examples of universities that, before 1800, taught that catastrophism accounted for the geologic column include (a) Yale; (b) Harvard; (c) Oxford; (d) Cambridge; (e) U.C. Berkley.

6. Examples of evidence for a young earth include: (a) Minimal oceanic sedimentary deposition; (b) The salinity of the oceans; (c) Fleshy tissue in dinosaur bones; (d) The deficit of helium in our atmosphere.

Matching

_____ 1. Just as natural events gradually transpire now, naturally occurring events have always gradually taken place in the past.

_____ 2. "All the fountains of the great deep were broken up"

_____ 3. "Where were you when I laid the foundations of the earth?"

_____ 4. "For since the fathers fell asleep, all things continue as they were from the beginning of creation."

_____ 5. "Let the waters under the heavens be gathered together into one place, and let the dry land appear."

_____ 6. The men who influenced Charles Darwin's thinking regarding uniformitarianism.

A.   Genesis 7:11

B.   Job 38

C.   2 Peter 3:3-6

D.   The doctrine of uniformitarianism

E.   Genesis 1:28

F.   Genesis 1:6-9

G.   Hutton and Lyell

True or False

1. ____   A 10,000 year age for our planet is called an "old-earth" view.

2. ____   If the great Noahic flood had actually occurred, then all types, kinds, and sizes of life forms would be interspersed throughout the geologic column.

3. ____   If the great Noahic flood had actually occurred, then we would not find similar layers of sedimentation in various parts of the world.

4. ___ The age of the earth is not relevant to the study of evolution.

5. ___ The Bible gives no indication as to what might have caused the geologic column.

6. ___ Adam, Eve, and all life forms were physically immature on the day that God created them.

Discussion Question in Preparation for Answering Unbelievers and Critics

You are eating dinner with a friend and her family. Her father, who is a meteorologist, begins talking about the age of the earth and comments, "It is not logical that God would make a world that appears to be billions of years old on the first day of creation. It takes millions of years of weathering to create the mature planet on which we live." He then asks you what you think of this idea. What do you say? _____

_____

_____

_____

_____

_____

_____

_____

_____

_____

_____

_____

_____

Lesson 12

The Ape/Man Question (The Origin of Man)

Introduction

Of all neo-Darwinian myths, there is none so widespread as the teaching that man evolved from ape-like, cave-dwelling ancestors. God has told us from where man came and by what means (Gen. 2). Atheistic agendas have long desired to discredit the biblical account. This lesson examines their methods.

I. Where's the Evidence?

Because this is the most disseminated of all neo-Darwinian dogmas, individuals far too often assume that there must be a wealth of evidence supporting this monkey to man myth. Further, because modern man is supposedly one of the last animal species to have evolved over the past few tens of thousands of years, and because his remains are the most sought after of all fossils, there should be *more* of them as well as less decayed fossil evidence in this area than in any other. But, there's not. An examination of the facts proves that this evolutionary hypothesis is one of the most deceptive, baseless, and

propped up of beliefs. In order to prove the type of macroevolution that would transform microbe to man there would of necessity be vast amounts of transitional fossils (discussed in Lesson nine) between the ape-like creatures that man allegedly descended from and modern man. Again, there's not.

In this scholarly work, two Ph.D. scientists "examines the fossil record, gender and sexual reproduction, language, the brain, mind and consciousness, skin color and blood types, etc. Each chapter presents an insurmountable obstacle for evolutionary theory, and discusses current scientific data that uphold the creation model ", as described by Apologetics Press. This 526-page book is available for purchase or FREE in PDF format from www.apologeticspress.org.

"The trouble is, we probably know more about the evolution of extinct trilobites than we do about the evolution of man" (Palmer, 2002, p. 50). And this is in spite of scientists admitting that the eyes of trilobite fossils exhibit no evolutionary change. (See Lesson ten regarding the research of Dr. Niles Eldredge.) As we will see, even the suggested fossil evidence for ape-man is highly questionable if not completely contrived. Far too often, researchers, in their haste to make a new discovery, will take a very few broken, shattered, and disconnected fossilized human bones, assign them ages of millions of years, reconstruct them into what ape-man supposedly looked like, and proceed to tell us how they walked, what they ate, and even what their facultative and cognitive abilities were. How can they know these things? As we will see... they can't.

II. Human Evolution: Science or Fiction?

The proposed mechanism for human evolution has and remains a voraciously held

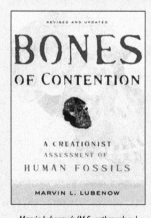

Marvin Lubenow's (M.S, anthropology) 400 page book undermines the human-evolution hypothesis as a "house of cards" exposing numerous frauds and faulty dating methods. (NOTE: Lubenow is an old-earth creationist).

From Answers in Genesis' Pocketguide Series, and described by them as, "Experts in the fields of paleontology, anatomy, genetics, and ancient Bible texts examine both the scientific evidence and the biblical record. They show that humans are not related to apes, but were specially created by God in His image." This is an excellent 96-page overview of the ape-man controversy.

A creationist's artistic rendering of the erroneous teaching of evolutionary transition from ape to man, commonly seen in many science text books. (Copyright Jody F. Sjogren 2000. Used with permission).

dogma, despite a lack of any evidentiary support. In fact, two major evolutionary camps even disagree as to where the evolution of the first man took place: in Africa, or simultaneously in various parts of the world. Even before supposed fossil proof had been unearthed, the evolution of man was already taught as a veritable fact. Charles Darwin, in his work, *The Descent of Man,* proposed, "the origin of man as a distinct species by descent from some lower form" although he readily admitted that no fossil evidence supporting this idea had ever been found! (1882, p. 613). However, Darwin set the stage for the elaborate scientific ruse that mankind's ancestral evolution could be proven by scientific means. This was not the case. Hear the Chief Science Writer for the prestigious science journal *Nature* admitting this fact in stating, "The intervals of time that separate fossils are so huge that we cannot say anything definite about their possible connection through ancestry and descent (Gee, 1999, p. 23).

The first Hungarian edition (1884) of Charles Darwin's second book, "The Descent of Man," where he argues that man evolved from ape-like ancestors.

In the same book (p. 127) Gee stated, "it is impossible when confronted with a fossil, to be certain whether it is your ancestor, or the ancestor of anything else, even another fossil. We also know that adaptive scenarios are simply justifications for particular arrangements of fossils made after the fact, and which rely for their justification on authority rather than on testable hypotheses."

He also stated, "No fossil is buried with its birth certificate. That, and the scarcity of fossils, means that it is effectively impossible to link fossils into chains of cause and effect in any valid way... In reality, the physical record of human evolution is more modest. Each fossil represents an isolated point, with no knowable connection to any other given fossil, and all float around in an overwhelming sea of gaps... as if the chain of ancestry and descent were a real object for our contemplation, and not what it really is: a completely human invention created after the fact, shaped to accord with human prejudices... To take a line of fossils and claim that they represent a lineage is not a scientific hypothesis that can be tested, but an assertion that carries the same validity as a bedtime story—amusing, perhaps even instructive, but not scientific" Gee, 1999, pp. 32, 113-117).

Elsewhere, Dr. Gee (an evolutionary paleontologist) admitted, "Fossil evidence of human evolutionary history is fragmentary and open to various interpretations. Fossil evidence of chimpanzee evolution is absent altogether" (Gee, 2001).

Chief Science Writer for the journal Nature, Dr. Henry Gee, admitted that to take a line of fossils, "and claim that they represent a lineage is not a scientific hypothesis that can be tested" (Gee, 1993, p. 23). (Photo: www.huffingtonpost.com).

In 2011, Dr. Gee was even bolder in admitting, "We have all seen the canonical parade of apes, each one becoming more human. We know that, as a depiction of evolution, this line-up is tosh. Yet we cling to it. ... Almost every time someone claims to have found a new species of hominin [ape-like ancestor], someone else refutes it. ... Dart's original paper on A. africanus was, it is true, long on waffle and short on substance" (Gee, 2011) Drs. Lowenstein and Zihlman (1988), in their article, "The Invisible Ape" corroborate Gee's sentiments in stating, "…anatomy and the fossil record cannot be relied upon for evolutionary lineages. Yet palaeontologists persist in doing just this." Dr. J. Shreeve in Discover magazine (1990) said, "Everybody knows fossils are fickle; bones will sing any song you want to hear."

Professor J.S. Jones, Department of Genetics and Biometry, University College, London, in the prestigious journal Nature admitted, "Palaeoanthropologists seem to make up for a lack of fossils with an excess of fury, and this must now be the only science in which it is still possible to become famous just by having an opinion. As one cynic says, in human palaeontology [the study of fossils] the consensus depends on who shouts loudest" (Jones, 1990). Notice the following two quotes from renowned evolutionists regarding the construction of human evolution from the fossil record, known as the science of paleoanthropology. The first quote is from Dr. Ian Tattersall, curator of the American Museum of Natural History and the second from Dr. Geoffery Clark, anthropologist at Arizona State University. "In paleoanthropology, the patterns we perceive are as likely to result from our unconscious mindsets as from the evidence itself" (Tattersall, 1996, pp. 47-54). Responding to Dr. Tatersall's quote, Dr. Clark said, "Paleoanthropology has the form but not the substance of science" (*Ibid.*). Dr. Bernard Wood, professor of human origins, George Washington University said, "There is a popular image of human evolution that you'll find all over the place... On the left of the picture there's an ape... On the right, a man...

Between the two is a succession of figures that become ever more like humans... Our progress from ape to human looks so smooth, so tidy. It's such a beguiling image that even the experts are loath to let it go. But it is an illusion" (Wood, 2002).

What we have, then, is not a science at all, but a long-held and deep-seated philosophy or belief system. One evolutionist, Misia Landau, addressed this in her 1991 work, Narratives of Human Evolution. Dr. Landau examined many classic evolutionary texts, illuminating that these evolutionary teachings were "determined as much by traditional narrative frameworks as by material evidence" (p. x). What does this mean? It means that the succession of human evolution, widely taught in text books, has been largely based on narrative, conjecture, story, folktale, and myth. According to Dr. Landau, four main events are repeated throughout these narratives: (1) human ancestors move from living in trees to living on the ground, (2) they develop an upright standing posture, (3) they acquire intelligence and language, and finally, (4) they develop technology and society. These are assumptions made with no scientific substantiation. Dr. Landau goes on to explain that modern descriptions of human evolution "far exceed what can be inferred from the study of fossils alone and in fact place a heavy burden of interpretation on the fossil record—a burden which is relieved by placing fossils into preexisting narrative structures" (p. 148). Dr. Landau is explaining that when someone questions this narrative (story) of human evolution, its proponents will refer the detractor to the fossil record as proof that

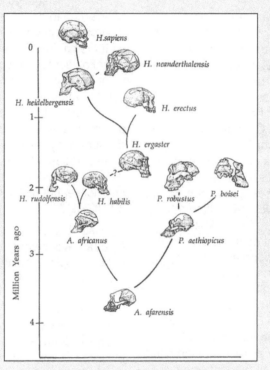

A Darwinist's diagram of the supposed descent of man. (Courtesy of Apologetics Press).

they should believe the evolution narrative. However, when one questions the paltry, incongruous, and unconvincing fossil evidence for human evolution, its proponents refer the detractor back to the evolution narrative as proof that they should believe the fossil record. This type of circular reasoning has no place in science. Remember "heads I win, tails you lose"? What occurs is molding the facts to fit the preconceived desired conclusions (i.e., "unconscious mindsets") of the popular scientific community, instead of the other way around.

Following Darwin's two landmark publications, there was a mad rush to find the human fossils that would serve as evolutionary transitional forms and validate Darwin's hypothesis; that is, finding the evidence to shape around, uphold, and authenticate the predetermined, presupposed idea of man's evolution.

III. Examining the Fossils

Time only permits us to briefly survey the supposed fossil evidence supporting the evolution of man. Please note that the burden of proof is on the evolutionist to provide the transitional missing links between man and ape. With regard to finding an original common ancestor from which man and ape branched off from and supposedly descended from some six million years ago, the August 23, 1999 issue of *Time* magazine reported the following, "Common Ancestor. No fossils found yet, but scientists believe the human and ape lines diverged." Evolutionists believe that the reason we do not see any half men/half apes continuing to evolve into man today is because, during the evolutionary process, the ape and human groups diverged and evolved along separate lines. Creationists have often failed to understand this point and, thus, have misguidedly questioned why there are no "half men, half apes" still evolving today.

A point should also be made regarding featural differences among humans. This can best be illustrated with the following analogy. Imagine that the skeletal remains of four very different structurally shaped groups of humans (who are coexisting on this planet today) are carefully buried in various locations on the earth. Imagine that these fossilized

remains are discovered 1,000 years from now. Future generations, having no knowledge of the current age we live in, and if inclined to believe in evolution, might erroneously conclude that these skeletal remains were from four different groups of animals. They might conclude that the four humans represent a line of evolution, one into the other. They would base their conclusions on differences in our height, brow ridge thickness, cranial capacity, flat or protruding facial features, and length, width, thickness, and density of bones. Today this type of variation may be seen among such diverse human groups as the African Mbuti pygmies who are less than five feet tall, the African Burundi Watusis who are often greater than seven feet tall, the Australian Aborigenes, Chinese Mongols, and Canadian Inuits. However, these physical differences only solidify our point that, despite variation within a kind, men have always been men and apes have always been apes. These future generations of evolutionists would be just as wrong to conclude that different men today are different creatures in various stages of evolution, as current neo-Darwinists are wrong when they draw these conclusions from our fossilized human ancestors.

A. Bona fide bogus bones. The first category of fossil evidence represents those findings that have long been proved deceptions, hoaxes, and fabrications, not the missing links necessary to prove human evolution.

1. Java Man (1887). One tooth, a piece of a skull and thigh bone were upheld as proof of the missing link between man and ape. The thigh bone was later admitted to be from modern man and the skull fragment from a gibbon (monkey).

Skull cap of a gibbon (ape) purported to be from "Java Man." (Courtesy of Apologetics Press).

2. Piltdown Man (1913). For forty years the jaw and skull fragments of Piltdown man were paraded before tens of thousands of museum visitors as the missing link. In 1953, it was proved that deception had been involved, as the Piltdown skull fragment was from a modern

Old bones, stained teeth, one trunk and the Missing Link

CHARLES ARTHUR
Science Correspondent

After 43 years of detective work, the search for the perpetrator of the biggest scientific hoax of the century is finally over – and the motive has been revealed as one man's wish for a weekly wage instead of piecework payment.

"Piltdown Man", a faked fossil discovered in 1912, ruined the reputation of Arthur Smith Woodward, keeper of palaeontology at the Natural History Museum. He went to his deathbed insisting that the skull discovered in a Sussex quarry was that of the earliest Englishman, and that the carved elephant bone found with it (shaped suspiciously like a cricket bat) was genuine.

But in 1953, five years after Woodward died, the fossils were shown to be fakes: the skull, instead of being the "missing link" between ape and man, was composed of an orangutan jaw and a man's head. The other fossils were also found to be fakes, made of stained and carved old bones.

However, the identity of the hoaxer remained a mystery. Over the years, it was blamed variously on Charles Dawson, a lawyer who first found the remains, on Sir Arthur Conan Doyle, creator of Sherlock Holmes, and on Teilhard de Chardin, the noted priest and palaeontologist.

But the discovery by the Natural History Museum of an old canvas trunk in its attic seems to have answered the question.

Martin A.C. Hinton (above, centre) is now thought to have created the skull of 'Piltdown Man' (below left)

Inside were human teeth, which had been stained like those of the "fossils". The trunk's owner was the late Martin A.C. Hinton, the museum's curator of zoology at the time of the fraud. "I'm 100 per cent certain that it was him," said Brian Gardiner, professor of palaeontology at King's College, London, yesterday. "The contents of the trunk clinch it."

Professor Gardiner first had a hunch that it was Hinton in 1945, when he was working at the museum as a student as the

fraud was exposed. He will give a lecture on his conclusion tomorrow night to the Linnean Society. "Hinton was known as a practical joker. Dawson was the fall guy for his practical jokes, just a gullible solicitor. Hinton's motive is shown by some letters," he said yesterday.

"In 1910, Hinton was just a summer student working there in his holidays, and he wrote to Woodward asking to work at the museum cataloguing rodent remains." He was offered £130 – after the work was complete. Hinton, then 27, asked for a weekly payment. Woodward is thought to have been unmoved – which piqued Hinton, a prodigy who at 16 had had a paper published on how fossils become stained by river deposits.

The contents of the trunk show that Hinton produced the fakes by careful staining: the teeth were his test run. The key clues for the Piltdown detectives are the presence of traces of chromium metal in the teeth, the trunk – and the Piltdown bones. The chromium is the missing link which finally fingers Hinton.

The only question that remains is why Hinton did not own up once Woodward had swallowed the bait. "I think it was all taken so seriously and attracted so much attention that he couldn't," said Henry Gee of the science journal Nature, which today publishes a full account of the search. "The trouble now is that all the suspects are dead and buried. You would have to be Inspector Morse to answer that one."

Article from the British newspaper, "The Independent" (Thurs. May 23, 1996), exposing the perpetrator behind the Piltdown hoax to be Arthur Smith Woodward, Keeper of Paleontology at the British Natural History Museum. Nevertheless, dozens of scientists must have been complicit with the fraud, as the filings and patchwork on the teeth were visible to the naked eye.

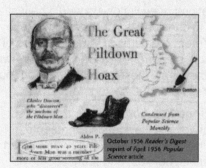

Reader's Digest exposing the Piltdown hoax in 1953.

man. The teeth in the jaw, from an orangutan, were filed down too far, a hole was patched with bone, and chemically treated to make them appear human. Although the perpetrator was identified in 1996, dozens of scientists must have been complicit in the fraud, as the filings and patchwork on the teeth were visible to the naked eye.

3. Rhodesian Man (1921). Fossils from three or four individuals were upheld as missing links, until years later when it was admitted that they belonged to modern man.

4. Nebraska Man (1922). From a single tooth, Nebraska Man was internationally proclaimed as the transition between ape and man along with reconstructed pictures such as this one from the June 24, 1922 *Illustrated London News*. Nebraska man was even entered as evidence in the 1925 Scopes "monkey trial." Two years later, however, this single tooth was admitted to be that of a wild pig. It has been said that this is the case of a pig making a monkey of a man.

5. Orce Man (1983). A piece of skull cap was purported to be from an evolutionary child in line to humans. Some time later it

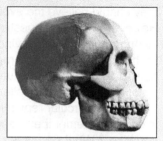

Area of modern human skull that was found and transformed into "Piltdown Man." (Courtesy of Apologetics Press).

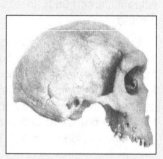

A human skull that was thought in 1921 to be from "Rhodesian Man." (Courtesy of Apologetics Press).

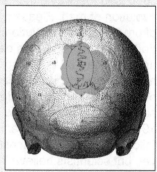

Skull cap area supposedly from "Orce Man," later shown to be part of a donkey's skull. (Courtesy of Apologetics Press).

was confessed that the skull fragment was actually that of a donkey.

6. Kenyanthropus platyops (2001). Dated 3.5 million years old and supposedly in line to humans, this late finding has been suggested as comprising a new genus. *K. platyops* was based, however, on questionable evidence: thirty-six cranio-dental fossils from four locations over seventeen

Illustration of "Nebraska Man" in The Illustrated London News, 24 June 1922, after a wild pig tooth (supposedly from a hominoid) was discovered in western Nebraska.

years, only six of which contained actual bone. Further damaging is the flat-faced features of the *K. platyops* that make it appear human. This is in stark contrast with other categories of fossils much younger than *K. platyops* that have more ape-like features and, thus, should be older according to evolutionary development (Harrub and Thompson, 2003, pp. 32-40).

B. Fossils shown to be those of apes. This second category includes those fossils previously thought and taught as being ancestral to man. Evolutionists attempt to explain these fossils in one of two ways. Either they admit these are the bones of genuine apes, or they skirt the issue saying they are neither apes nor in line to humans, but compose a "side branch" of evolutionary creatures that are now extinct.

1. "Lucy"—*Australopithecus afarensis*, (1974). Many scientists still profess Lucy to be either an evolutionary side branch to man or in line to man, despite the wealth of evidence to the contrary. However, other evolutionists profess, as creationists do, that Lucy was nothing more than an ape. Admitted proof for this position includes Lucy's stiff wrist, which indicates that she walked on all fours on her knuckles as an ape, not upright like modern man, as evolutionists claim. Other evidence includes chimp-proportioned arm bones, and chimp-like thumbs, skull, and brain. The world renowned journal *Science* even reported that Lucy was a side branch not in line to humans (see Shreeve, 1996, p. 654).

2. *Ardipethecus ramidus kadabba* (1994). *A. ramidus* was the creature named from the findings of a fragment of jaw bone and two hand bones, one collar bone fragment, three arm bones, one foot bone, and a few teeth gathered from five locations over five years. By all appearances, *A. ramidus* was an ape. First dated at 3.8 million years, the bones were quickly re-dated at up to 5.8 million years. Detractors include Donald Johanson,

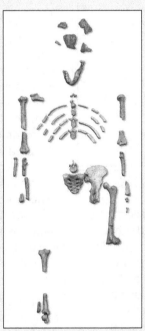

The handful of bones discovered by Donald Johansen in 1974 comprising "Lucy" (Australopithecus afarensis), shown NOT to be an ancestor of modern man. (Photo credit: Von 120 (2007))

discoverer of Lucy, who said in the *National Geographic* that *A. ramidus* had "many chimp-like features" and that "its position on the human family tree is in question" (1996, p. 117). Other scientists in *Time* magazine have stated that the discoverers of the fossils, "haven't collected enough bones yet to reconstruct with precision what kadabba looked like" (Lemonick and Dorfman, 2001).

C. Fossils shown to be those of mere modern mortal man. This third category includes those fossils previously thought to be evolutionarily ancestral to man, but have now been shown to be those of modern man.

1. *Homo habilis/Homo rudolfensis.* This genus has been called a wastebasket taxon and a dumping ground for fossils with no home. The group is not clearly defined, represents no missing link between ape and man, and its members have even been shown to have been builders of circular stone shaped huts, a human trait. Additionally, *Homo habilis* has been demonstrated to have coexisted along side of *Australopithecus* whom *Homo habilis* supposedly descended from up to 2.6 million years earlier! (Lemonick and Dorfman, 1999).

2. *Homo erectus/Homo Ergaster.* These two have been recently relegated into one group and, again, there is nothing to suggest that these individuals were any more or any less than human. Some argue that proof for *Homo erectus* being a missing link is that his cranial capacity was 850 to 1100 milliliters—very small. However, this is well within the range of humans today, who have cranial capacities of between 700 and 2200 milliliters; such as the African Mbuti pygmies who are less than five feet tall. Further, evolutionists teach that *Homo erectus* and modern man coexisted and that the youngest *Homo erectus* fossil is dated at 12,000 years old, well within the range of modern man (Lubenow, 1992, pp. 120, 127, 129).

THEOPHILUS   *Discovering Nebraska Man*

I've found it!
I've found a missing link!

The owner of this tooth was ape-like, tall, walked erect, had deep set eyes and projecting eyebrows.

Let me see that

No— I'd say the owner is short, walks on all four hoofs, has big ears, a long mobile snout and curly tail.

www.theophilus.org

©BOB WEST 1997

Rights owned and administered by Bob West. Used with permission.

3. Neanderthal Man—*Homo neanderthalis*
(1856). Of all groups examined, Neanderthal man most clearly represents modern man with physical variation. Many evolutionists have recently come to the same conclusion as that of the Neanderthal man discoverer Rudolph Virchow, who believed that Neanderthal's bones were from modern rather than from a prehistoric man. The fossils, first discovered in the Neander Valley of Germany, represent a group of individuals thought to have rickets and osteomolacia, conditions brought on by a vitamin D deficiency, due to lack of exposure to sunlight. The disease manifests itself in weakened and calcium deficient bone tissue that leads to many skeletal deformities including elongated and bowed leg bones, excessive cranial growth creating enlarged and thickened skulls, and enlarged, protruding brows. For those who are willing to overlook the facts and are inclined to prove human evolution, *Homo neanderthalis* represents a boorish, crude, dim-witted beast side chain in line to modern man. This is in spite of evolutionary teaching that Neanderthal man, like *Homo erectus,* coexisted side by side with modern man for up to 40,000 years! Could it be that Neanderthal man was simply a modern man

with a slightly varied physical appearance? Why should we believe otherwise? Scientists further admit that Neanderthal man engaged in all the habits, customs, and nuances of modern man, including shelter building, tool utilization, the creation of jewelry, art, and five different musical instruments, and the ritual burial of the dead, some alongside of what are considered to be modern man. See Appendix G for "Closing the Door on the Non-Human Neanderthal Myth."

IV. The Truth about the Origin of Man

As was mentioned in earlier lessons, God and His inspired word must be our ultimate source of authority. Moses, in Genesis 2:7, tells us that man was formed on the sixth day of creation in the following manner, "And the Lord God formed man of the dust of the ground and breathed into his nostrils the breath of life; and man became a living being." This man, Adam, was further verified as the first man by Paul, "And so it is written, 'The first man Adam became a living being'" (1 Cor. 15:45). To corroborate the fact that Adam was the first man created at the outset of the world, Jesus said, "But from the beginning of creation, God 'made them male and female'" (Mark 10:6).

Some Christians teach that Adam was only a metaphor, an allegory for the primitive evolutionary stages of man. Or, they teach that Adam himself was an unevolved cave-man whom God created to christen the human evolutionary process. One religious teacher said, "Man has been in a constant state of evolution" and "This writer sees no need to view Adam as a highly advanced and sophisticated individual. God had to make the first clothes man wore so he wasn't very advanced" (Clayton, 1976, p. 133, Clayton, 1978). Assuming Adam to be an unsophisticated, unadvanced, bumbling cave dweller is a spurious theological argument that overlooks two very important facts. (1) The receipt of a gift (i.e., Adam's clothes) does not indicate innate ignorance on the behalf of the receiver, and (2) Adam and his immediate ancestors displayed numerous advanced traits of intelligence that separated him from mere animals (Table 1, on the facing page).

Table 1. Examples of the advanced intellectual capacities of the first humans in the Bible

Individuals	Scripture in Genesis	Indication of Intellectual Ability
All	1:27	Man's creation in the *"image of God"*
All	1:26, 28	The ability to *"subdue"* and *"have dominion… over all the earth and over every creeping thing that creeps on the earth"*
Adam and Eve	1:28-31	The ability to comprehend and process intelligible questions
Adam	2:15	The ability to tend and keep a garden of flowering plants and animals
Adam	2:19-20	Naming all the cattle, birds of the sky, and beasts of the field
Adam	2:23; 3:20	Naming another human, Eve
Adam and Eve	2:25; 3:7, 8	The ability to be ashamed

The creation of the first man by God *ex nihlo* (from nothing) was a miraculous event that resulted in a race of soul-filled human beings who possessed equal or superior intellectual ability to mankind today.

NOTES
_____
_____
_____
_____
_____
_____

Questions

Matching

_____ 1. Rickets

_____ 2. Nebraska man

_____ 3. Event # 1 in the fabricated human standing evolutionary narrative

_____ 4. Event # 2 in the fabricated human evolutionary narrative

_____ 5. Event # 3 in the fabricated human evolutionary narrative

_____ 6. Event # 4 in the fabricated human evolutionary narrative

a. Human ancestors develop technology

b. The case of a pig making a monkey out of a man

c. Human ancestors develop an upright posture

d. A disease caused by vitamin D deficiency leading to the elongation, bowing, and thickening of the human bones that may be found in the fossils of the people of the Neander valley of Germany

e. Human ancestors move from living in trees to living on the ground

f. Human ancestors develop language

True or False

_____ 1. The monkey to man belief is not widely held among evolutionists.

_____ 2. The study of the evolution of man is a philosophy rather than a science.

_____ 3. Piltdown Man was a hoax, created from filed down and chemically treated bones.

_____ 4. "Lucy" had flexible wrists, like humans.

_____ 5. Some Christians deny that Adam was actually the first man created. Instead, they teach that he was the result of evolution.

Short Answer

1. Name some habits of Neanderthal man that suggest that he was human. _____
_____
_____

2. Which three prophets declared that Adam was the first man? _____
_____
_____

3. What evidence do we have that the first people created in Genesis 1-4 were highly intelligent human beings? _____
_____
_____
_____

4. Name the three categories in which supposed "ape-man" fossils may be placed. _____
_____
_____
_____

5. Who was the author that made famous the monkey to man hypothesis in his 1871 publication, *The Decent of Man,* and what evidence did he provide? _____

_____

Discussion Question in Preparation for Answering Unbelievers and Critics

You are tutoring two fifth-grade students in life science. They are studying various classes of animals. You teach them that, as humans, we are classified as mammals. At this point, one child says to the other, "I knew that we were mammals because I've seen bones in a museum that prove that we were apes before we evolved into humans." What do you say?_____

_____

_____

_____

_____

Lesson 13

The Social Fruits of Evolutionary Teaching

Introduction

Is there danger in a society rejecting the truth regarding the origin of mankind and our eternal destiny? What if a culture accepts neo-Darwinian dogmas of humanism, uniformitarianism, and naturalistic evolution? Will there be any irreparable damage? This lesson will detail the consequences of this dark and destructive doctrine that will ultimately lead to theft, brutality, murder, rape, unwanted pregnancies, infanticide, genocide, sexually transmitted diseases, and rank immorality in general. This is not to say that evolutionary propaganda single-handedly creates all of these problems. However, evolutionary teaching has certainly exacerbated these maladies far beyond what they would be otherwise.

I. Evolutionary Teaching Promotes Atheism

Recall that the premise behind the creation of the evolutionary hypothesis was to destroy belief in God. Witness the stunning quotes that follow by professed evolutionists. "For what religious man came eventually to think of as 'conscience' is simply the faculty that enabled his hominid ancestors to inhibit their programmed responses to stimuli" (Allegro, 1986, p. 26). "Human beings are the natural culmination of millions of years of evolution.... No educated man or woman can possibly believe in the Christian notion of bodily resurrection" (Lamont, 1991, pp. 5-8). "Religion is compatible with modern evolutionary biology... if the religion is effectively indistinguishable from atheism" (as quoted in Provine, 1987, p. 51). D.M.S Watson stated, "... evolution itself is accepted by zoologists, not because it has been observed to occur... but because the only alternative, special creation, is incredible" (1929, p. 233). Another evolutionist observed, "... From the earliest stages of Greek thought man has been eager to discover some natural cause of evolution, and to abandon the idea of supernatural intervention in the order of nature" (Osborn, 1918, p. ix). Another atheist admitted, "Evolution thus becomes the most potent weapon for destroying the Christian faith" (Matthill, 1982). Sir Julian Huxley, stated, "Darwinism removed the whole idea of God as the creator. . .Darwin pointed out that no supernatural designer was needed; since natural selection could account for any known form of life, there was no room

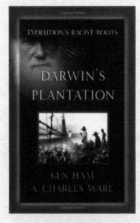

Ken Ham and Dr. Charles Ware, in Darwin's Plantation, explore the history of Darwin's racist roots and how his teaching in The Descent of Man demeaned darker colored peoples as "savage races," encouraging the extermination of thousands of people around the world (2007).

Dr. Richard Weikart, professor of history at California State University, and Senior Fellow of the Discovery Institute, wrote From Darwin to Hitler: Evolutionary Ethics, Eugenics, and Racism in Germany, to expose the revolutionary impact Darwinism had on ethics and morality throughout history (2006).

In Dr. Richard Weikart's follow-up book, he explains how "Hitler was inspired by evolutionary theory to pursue the utopian project of biologically improving the human race, and this ethic underlayed or influenced almost every major feature of Nazi policy" (2011).

for a supernatural agency" (Huxley, 1945, p. 45). Another stated, "Man is the result of a purposeless and materialistic process that did not have him in mind... the workings of the universe cannot provide any automatic, universal, eternal, or absolute ethical criteria of right and wrong" (Simpson, 1951, p. 179, 180). Woolsey Teller, one time president of The American Association for the Advancement of Atheism, stated, "Evolution produced the microbe as well as man. What intelligence is there in that?... Evolution means Atheism... Evolution spells Atheism—a godless universe. . .The God idea cannot be reconciled with our knowledge of evolution" (Teller, 1945). Others have stated, "In the evolutionary pattern of thought there is no longer need or room for the supernatural" (Huxley, 1960, pp. 252, 253). "Darwinism and Neo-Darwinism, rightly or wrongly, have been used everywhere... as the main weapon against the biblical doctrine of origins" (Wilder-Smith, 1975, p. 31).

Harvard entomologist and avowed atheist, Dr. E.O. Wilson (Harrison).

In 2003, I was able to meet and converse with Dr. E.O. Wilson (Harvard professor, renowned evolutionist, and two-time Pulitzer prize winner) on the University of Georgia campus. Dr. Wilson personally told me that Darwinism was the reason he had abandoned his conservative religious faith and upbringing. Here is his account of this event, which I found later, "As were many persons from Alabama, I was a born-again Christian. When I was fifteen, I entered the Southern Baptist Church with great fervor and interest in the fundamentalist religion; I left at seventeen when I got to the University of Alabama and heard about evolutionary theory" (1982, p. 40).

II. Evolutionary Doctrine Destroys Morality

Aldous Huxley, atheist and evolutionist, admitted that his motivation to believe naturalism was rooted in his desire to eliminate morality: "I had motives for not wanting the world to have meaning... The liberation we desired was simultaneously liberation from a certain political and economic system and liberation from a certain system of morality. We objected to the morality because it interfered with our sexual freedom" (1946, pp. 270-273). Others have stated, "Morality needs no supernatural sanctions or motivations. It is not heaven-sent; it is human-evolved" (Morain and Reiser, 1988). Famed evolutionist Richard Dawkins (known as "Darwin's pit bull" or "Darwin's Rottweiler," which is a play on T.H. Huxley's nickname, "Darwin's Bulldog"), in regard to evolution stated, "You are for nothing. You are here to propagate your selfish genes. There is no higher purpose in life." Ironically, in the same book he admits "... a human society based simply on the gene's law of universal ruthless selfishness would be a very nasty society in which to live" (1989, pp. 2, 3).

And with those words, Dawkins makes our point most eloquently. When people believe and practice evolutionary tenets, our world becomes increasingly more immoral, ruthless, selfish, dangerous, and self-destructive. And is it any wonder? When the idea of God is thoroughly discarded, what is there that constrains man to adhere to any principles of morality or decency? The French Philosopher Jean Paul Sartre said, "Everything is indeed permitted if God does not exist... Nor, on the other hand, if God does not exist are we provided with any values or commands that could legitimize our behavior" (as quoted in Marsak, 1961, p. 279). The *Humanist Manifesto*, drafted to destroy deity in light of evolution, even promotes immoral behavior, in stating, "... We believe that intolerant attitudes, often cultivated by orthodox religions, and puritanical cultures, unduly repress sexual conduct. The right to birth control, abortion, and divorce should be recognized" (1973, pp. 18, 19). Due to the naturalistic belief, that we exist as a result of a purposeless, hopeless, hapless, loveless, random, cosmic accident apart from a divine Creator, man is led into a series of self-destructive conclusions, reeking havoc on himself and all those around him.

III. Evolutionary Teaching Promotes Abortion

Those who adhere to a belief in God and biblical standards hold human life as sacred and believe murder to be iniquitous. Over 3,400 years ago Moses, through the Holy Spirit, dictated the gravity of this principle: "Whoever sheds man's blood, by man his blood shall be shed; for in the image of God He made man" (Gen. 9:6). Darwinism, however, dispenses with this principle replacing it with *"survival of the fittest."* This teaching led two renowned evolutionists to compare the human embryo to embryos of any other animal. They taught that because the developmental stages of the human infant are comparable to evolutionary change from amoeba to fish

The late evolutionist Carl Sagan compared aborting a fetus to stepping on a toad. (Photo by Michael Okoniewski, Copyright © 1994. Used with permission).

to amphibian to reptile, then aborting a human fetus is no different than *"stepping on a toad"* (Sagan and Druyan, 1990). Further, neo-Darwinian theory states that weaker, feebler, more dependent creatures should be disposed of to make way for the stronger. Unwanted human fetuses would certainly meet this criterion.

Outspoken supporter of macroevolution, Dr. Peter Singer, Professor of Bioethics at Princeton University, who believes that parents should be allowed twenty-eight days after the birth of their child to decide whether it should live or be euthanized. (Photo: Singer, 2009).

Dr. Peter Singer, Professor of Bioethics at Princeton University, outspoken evolutionist (and father of the modern animal rights activist movement) said, "Human babies are not born self aware, or ca-

pable of knowing that they exist over time. They are not persons... The life of a newborn is of less value than the life of a pig a dog or a chimpanzee" (Singer, 2011) and "A period of twenty-eight days after birth might be allowed before an infant is accepted as having the same right to live as others" (Kuhse and Singer, 1986), and "Newborn human babies have no sense of their existence over time. So killing a newborn baby is never equivalent to killing a person, that is, a being who wants to go on living" (Singer, 2015).

IV. Evolutionary Teaching Promotes Euthanasia and Eugenics

Why stop with unwanted pregnancies? What about the mentally ill, sick, elderly, blind, deaf, handicapped, and physically disabled? After all, wild animals in any of these conditions would certainly not survive, would they? Adolph Hitler came to this very conclusion, and before instituting racial genocide he instituted *"Aktion 14f13,"* which emptied the insane asylums in Germany and exterminated over 100,000 human beings considered "life unworthy of living," including the mentally impaired, handicapped, sick and disabled (Goldenhagen, 1997, pp. 119, 520). One Nazi propaganda film, decrying the mentally retarded said, "All weak living things will inevitably perish in nature. In the last few decades, mankind has sinned frightfully against the law of natural selection. We haven't just maintained life unworthy of life, we have even allowed

When Adolf Hitler wrote his treatise Mein Kampf (or My Struggle), he drew from the teaching of Charles Darwin to create his doctrine of the extermination of inferior races and those considered to be "life unworthy of living."

A 1939 Nazi propaganda poster, designed to convince the German public that the mentally retarded and disabled should be euthanized. The translation is, "It costs 60,000 Reichsmarks to look after this mental patient per year. Citizens—this is your money too!

Mentally ill inmates at Buchenvald camp designated to be exterminated under Hitler's Aktion14f13.

Transfer of mentally ill patients from Liebenau hospital on October 2, 1940, to be euthanized.

God's precious children at Schönbrunn Psychiatric Hospital for the mentally ill in 1934. Considered, "life unworthy of living," they were later euthanized under Hitler's Aktion14f13.

it to multiply! The descendants of these sick people look... like this person here!" (*Victims of the Past*, 1937). Evolutionists may disparage the Führer's unholy conduct, but they cannot deny that these are the inevitable fruits of their teaching. They would like to think of themselves as more civilized than a barbarian like Hitler by avowing a belief in love, charity, generosity, selflessness, and mercy to those less fortunate. However, if we are a product of nature rather than divine creation, all of these virtues are irrelevant. In fact, none of these virtues is anywhere found in Darwinism, but in God's word which teaches mercy toward those less fortunate (Luke 10:25-37). (As a side note, was Hitler's extermination of people considered "life unworthy of living" any different than what Americans do today when they abort their infants after genetic counseling reveals that the child will be born with Down Syndrome?)

Even Stephen J. Gould admitted that in order for evolution to be successful, "The price of perfect design is messy relentless slaughter" (Gould, 1990). The term *eugenics* was developed by Charles Darwin's half-cousin Sir Francis Galton, who is known as the "Father of Eugenics." He was mesmerized by Darwin's theory of evolution and believed (as did Darwin) we could improve the genetics of mankind by eliminating the "savage races." He said, "There is perhaps some connection between this obscure action and the disappearance of most savage races when brought into contact with high civilization... But while most barbarous races disappear, some, like the negro, do not... What nature does blindly, slowly and ruthlessly, man may do providently, quickly, and kindly. As it lies within his power, so it becomes his duty" (Galton, 1904). Commenting on Galton's

influence, Dr. Richard Brookes said, "Since Darwin's death, all has not been rosy in the evolutionary garden... A direct line runs from Darwin, through the founder of the eugenics movement—Darwin's cousin, Francis Galton—to the extermination camps of Nazi Europe" (Brookes, 1999).

V. Evolutionary Teaching Promotes Corrupt Philosophies

One of the most damaging doctrines of the twentieth century was socialist Marxism that gave rise to communism. When Karl Marx wrote his book *Das Capital,* he drew heavily from Charles Darwin's teachings and even dedicated his book to him. This led to Soviet communism, which subjected over fifteen countries to forty-five years of humanistic teaching that denied the existence of God and heralded the accomplishments of mankind

Karl Marx leaned on the teachings of Charles Dawin to construct his atheistic/socialistic doctrines that led to 20th century communism.

Sir Francis Galton was mesmerized by his second-cousin, Charles Darwin's, idea of human evolution and believed we could improve the gene pool by eliminating the "savage races."

The end result of Darwinism is the belief that our life is a purposeless waste. (Rights owned and administered by Bob West. Used with permission).

apart from the divine. I witnessed the fruits of this atheistic propaganda in the former Soviet Union firsthand in 1996 and 1997 in people who had never held a Bible and/or were brainwashed into believing there was no God. Is it any wonder that by 1991 the whole Soviet system collapsed on itself? Again, evolutionists may decry such a political system, but they cannot deny that Darwinism was the philosophy upon which the communist Soviet Union was founded. Governments that banish God and morality from society are certainly destined for shame, as the Proverb writer warned, "Righteousness exalts a nation but sin is a reproach to any people" (Prov. 14:34).

VI. Evolutionary Teaching Promotes the Elimination of "Genetically Inferior" Individuals

If, according to evolution, it is proper for the strongest and fittest to eliminate the weakest and most inferior, then Adolf Hitler's idea of cleansing the gene pool of the "*Untermenschen*" (subhumans) was absolutely commendable! Hitler adhered very closely to the Darwinian principle of *"survival of the fittest."* His idea was to help the world become a stronger, better place to live with a race of super people. Hitler was simply trying to help the evolutionary process along. In a book entitled *Hitler's Personal Security,* P. Hoffman wrote, "Hitler believed in struggle as a neo-Darwinian principle of human life that forced every people to try to dominate all others; without struggle they would rot and perish. ...Even in his own defeat in April 1945, Hitler expressed his faith in the survival of the stronger and declared the Slavic peoples to have proven themselves the

stronger" (1979, p. 264). The evolutionist Sir Arthur Keith earlier stated, "To see evolutionary measures and tribal morality... we must turn again to Germany of 1942. We see Hitler devoutly convinced that evolution produces the only real basis for a national policy. ...The means he adopted to secure the destiny of his race and people were organized slaughter, which has drenched Europe in blood... Germany has reverted to the tribal past, and is demonstrating to the world, in their naked ferocity, the methods of evolution" (Keith, 1979, p. 28).

In a speech Hitler made in Nuremburg in 1933, he said, "The Germans were the higher race, destined for a glorious evolutionary future. For this reason it was essential that the Jews should be segregated, otherwise mixed marriages would take place. Were this to happen, all nature's efforts to establish an evolutionary higher stage of being may thus be rendered futile" (Jappah, 2007)

Professor Jerry Bergman said, "A review of the writings of Hitler and contemporary German biologists finds that Darwin's theory and writings had a major influence upon Nazi policies. Hitler believed that the human gene pool could be improved by selective breeding... In the formulation of his racial policies, he relied heavily upon the Darwinian evolution model, especially the elaborations by Spencer and Haeckel. They culminated in the 'final solution,' the extermination of approximately six million Jews and four million other people who belonged to what German scientists judged were 'inferior races' " (Bergman, 1992).

Hitler and his entourage, including Herman Goering, head of the German Air Force, until their deaths predicted that history would one day vindicate their actions. Goering prophesied that one day mankind would praise them for their contribution to bettering the human race by eliminating the inferior. And, if evolution is fact, then Nazi Germany did its best to promote the process and cannot be condemned. God's word, on the other hand, teaches that all men are created equal and should be treated as such, without partiality (Acts 10:34; Rom. 2:11; Jas. 2:1-5).

Contrast this with Christian behavior. One non-Christian historian admits that it was the evangelicals that led the anti-slavery movement, as follows, "It was not the philosophies of Paris or Edinburgh or East Prussia who fought slavery, but the evangelical Christians and Quakers who drew their inspiration not from philosophy but from "superstitious religion". It was from the Evangelical Revival that the loudest claims for what we now call racial equality came" (Kenny, 2007).

VII. Evolutionary Teaching Promotes Unethical Business Dealings

The corrupt corporate world is presently immersed in scandal in the form of greed, deception, theft, and extortion. Is it any wonder that this would occur, considering the indoctrination of evolutionary principles that teach that the strongest and most ruthless should prevail, while the weak and naïve are to be trampled underfoot? Have you ever heard the expression, "The good guy finishes last"? The National Bank of Australia once cited the neo-Darwinian doctrine of *"survival of the fittest"* as the reason for their tremendous growth and success (Ham, 1987, p. 90). Thus, today's stockholders are swindled out of their life's savings, the naïve are deceived into accepting balloon loans with exorbitant interest rates, making repayment impossible, cyber-crooks pose as the needy to attain access to unsuspecting philanthropists' bank accounts, and middle-class working Americans are conned into investing into bogus pyramid schemes only to lose all of their hard earned-dollars when the company skips town. In short, individuals losing their scruples in order to profit at their neighbor's expense. But, isn't that what evolution teaches? Of course it is! The stronger, crueler, most devious, malicious, ravenous, and cannibalistic, according to Darwin, should benefit at the demise of the weak. God, on the other hand, teaches just the opposite and demands a higher standard of ethical integrity in our business dealings (Matt. 18:23-33; 7:12; Ps. 15:5; Prov. 14:31).

VIII. Darwin's Writings Promote Racism

In fact, his book, *On the Origin of Species by Natural Selection* was subtitled" *The Preservation of Favoured Races in the Struggle for Life,"*

THE ORIGIN OF SPECIES
BY MEANS OF NATURAL SELECTION
OR
THE PRESERVATION OF FAVOURED RACES IN THE STRUGGLE FOR LIFE

expressing his racist views. In his second book, Darwin stated, "At some future period, not very distant as measured by centuries, the civilized races of man will almost certainly exterminate and replace the savage races throughout the world. At the same time the anthropomorphous apes...

Darwinian supporter, Edward Ramsay, curator of the Australian Museum in Sydney had the Aborigines people killed and placed in the museum to represent the "missing link" in human evolution. He published a booklet on how to plug bullet holes in freshly killed "specimens." Ramsay also requested skulls of the "Bungee Blacks." Four weeks later a scientist sent him two corpses, proudly proclaiming, "The last of their tribe, had just been shot."

will no doubt be exterminated" (1874). In a letter to a friend, Darwin stated, "Looking to the world at no very distant date, what an endless number of the lower races will have been eliminated by the higher civilized races throughout the world" (1881). Regarding the human race, Darwin taught that blacks descended from the strong but unintelligent gorillas, the Orientals from the orangutans, and the whites from the most intelligent chimpanzees.

As a result of Darwin's teaching, Australian Aborigines specimens were in high demand to prove they were the missing link in human evolution. Published documents prove that more than 10,000 Aboriginal remains were shipped to British museums. The Smithsonian Institute still holds the remains of more than 15,000 Aborigines. When there were only four remaining Aborigines on the Island of Tasmania, Darwin requested their skulls, provided that the request did not "upset" their feelings. Many museums wanted the skins of Aborigines or freshly killed specimens so they could be stuffed. Dr. Edward Ramsay, curator of the Australian Museum in Sydney, published a booklet on how to plug bullet holes in freshly killed "specimens." Ramsay also requested skulls of the "Bungee Blacks." Four weeks later a scientist sent him two corpses, proudly proclaiming, "The last of their tribe, had just been shot." Native Americans were displayed as "emblematic savages" in the anthropology wing of 1904 World's Fair in St. Louis with Africans. Scientists severed one of the African's heads and boiled off the flesh so they could examine the skull. They were surprised to find that the brain capacity was larger than the statesman, Daniel Webster (Hallet, 1973). Evolutionist Dr. Stephen J. Gould of Harvard explained, "Biological arguments for racism may have been common before 1859, but they increased by orders of magnitude following the acceptance of evolutionary theory" (Gould, 1977).

IX. Evolutionary Teaching Promotes Abusive Male Supremacy and the Exploitation of Females

For centuries, women have been mistreated and denied basic human rights because of their status in society. This in spite of Jesus' teaching and example to the contrary (Matt. 8:14, 15; Mark 5:35-42; Luke 7:36-50; 23:27-31; John 8:3-11; 19:26, 27). About the time that Darwinism first took root, women's suffrage (the women's rights movement, not to be confused with twentieth century feminism) was also beginning to take a stand in developed societies. However, the two philosophies were diametrically opposed, as the suffrage movement was based on the principal of human equality, whereas Darwinism taught that the physically stronger of the sexes should prevail in oppression of the weaker. Eveleen Richards commented on this paradox: "In a period when women were beginning to demand the suffrage, higher education and entrance to middle-class professions, it was comforting to know that women could never outstrip men; the new Darwinism scientifically guaranteed it... an evolutionary reconstruction that centers on the aggressive, territorial, hunting male and relegates the female to...

the periphery of the evolutionary process" (1983, p. 887). Affirming the evolutionary superiority of man over woman, Charles Darwin stated, "It is generally admitted that with woman the powers of intuition, of rapid perception, and perhaps of imitation, are more strongly marked than in man; but some, at least, of these faculties are characteristic of the lower races, and therefore of a past and lower state of civilization. The chief distinction in the intellectual powers of the two sexes is shown by man attaining to a higher eminence in whatever he takes up, than woman can attain—whether requiring deep thought, reason, or imagination, or merely the use of the senses and hands" (Darwin, 1872b).

Male chauvinists were comforted by and latched onto the Darwinian doctrine of female inferiority. God's word, on the other hand, although placing woman in subjection to her husband, makes her equal in personhood, while distinct in work (1 Tim. 2:13-15). Additionally, Jehovah teaches that love, respect, honor, and admiration should be shown to the weaker vessel, in contrast to the necessary Darwinian conclusion of brutal male domination (Eph. 5:24, 25; 1 Pet. 3:7; Prov. 31:11, 28, 29). "Nevertheless, neither is man independent of woman, nor woman independent of man, in the Lord. For as woman came from man, even so man also comes through woman; but all things are from God" (1 Cor. 11:11, 12).

X. Evolutionary Teaching Promotes Scientific Wastefulness

Untold millions of dollars have been spent attempting to prove the evolutionary hypothesis. In addition to never being proven, no lasting good is accomplished through this type of research, with the exception of an occasional residual benefit. By and large, dollars spent on evolutionary experimentation is a gargantuan misappropriation. Indeed, much good is gleaned through valid scientific exploration such as cures for cancer and other debilitating diseases, increasing the proficiency of food production and battling the threat of pathogenic microorganisms. This, God's word encourages us to do (Gen. 1:26, 28). This has not and will not occur through research conducted in order to support a preconceived, yet unprovable philosophy.

Avraham Sonenthal said, "No product, discovery, medical procedure, or advance has come out of evolutionary theory. Without evolutionary theory, all practical biology would stand just as it is. No major corporation has a 'Department of Evolution' because scientists who have to produce results don't use it... In fact, I would like to challenge the readership of this publication to come up with one practical application of biology that would have been impossible were it not for the hypothesis of evolution" (Sonenthal, 1997). Professor Philip Skell, the "father of carbene chemistry" said, "Certainly, my own research with antibiotics during World War II received no guidance from insights provided by Darwinian evolution. Nor did Alexander Fleming's discovery of bacterial inhibition by penicillin. I recently asked more than seventy eminent researchers if they would have done their work differently if they had thought Darwin's theory was wrong. The responses were all the same: No.... I found that Darwin's theory had provided no discernible guidance, but was brought in, after the breakthroughs, as an interesting narrative gloss" (Skell, 2005).

Conclusion

The religion of evolution has been, is, and will continue to be a blight upon the well-being of society and the eternal destiny of man, resulting in irrevocable and everlasting damage. As long as righteous men and women are silent in the face of its conspicuous lies, it will continue to flourish. We should make a stand, "casting down arguments and every high thing that exalts itself against the knowledge of God, bringing every thought into captivity to the obedience of Christ" (2 Cor. 10:5); "avoiding worldly and empty chatter and the opposing arguments of what is falsely called 'knowledge'—which some have professed and thus gone astray from the faith" (1 Tim. 6:20, 21, NASB).

Upon our faith in Jesus as God become flesh, and His death, burial, and resurrection, we may find the freedom not offered by this world; but, in the Christ alone (Acts 4:12). This salvation is attained by laying aside our old lawless ways that Jesus calls sin (Rom. 3:23; 6:23), confessing Jesus as Lord of all (Phil. 2:9-11), and putting on Jesus as a garment through baptism (Gal. 3:27). Only through baptism into Jesus may we become His disciples (Matt. 28:19), have remission of our sins (Acts 2:38), be saved (Mark 16:16; 1 Pet. 3:21), have the right to go on our way with joy (Acts 8:26-39), have our sins washed away (Acts 22:16), be buried with Christ

The Holy Bible, "to the Jews a stumbling block and to the Greeks foolishness" (1 Cor. 1:23). (Photograph by Charles McCown courtesy of Apologetics Press).

and raised to walk in newness of life (Rom. 6:3, 4), and attain the circumcision of the heart (Col. 2:11, 12). Only those faithful to Him till death will attain the right to partake of the tree of life for an unfathomable eternity (Rev. 22:1-5). The alternative is unimaginable (Matt. 25:41; Rev. 21:8).

NOTES

_____
_____
_____
_____
_____
_____

Questions
TRUE OR FALSE

_____ 1. Some scientists resort to believing in evolution because they refuse to believe in a divine creator.

_____ 2. Over 3,400 years ago Noah said, "Whoever sheds man's blood, by man his blood shall be shed."

_____ 3. All criminals resort to evil-doing as a result of being taught evolution.

_____ 4. If evolution is truly followed to its unavoidable end, it will lead to a world where people of higher learning and deeper understanding correct the problems in society and create a harmonious existence of financial and social equality known as socialism.

_____ 5. God doesn't want us to achieve success in our business dealings, if it means destroying others in the process, or being dishonest with others to get to where we want to be.

_____ 6. God did not intend for women to be in subjection to their husbands.

Fill in the Blank

1. _____ _____ drew from Darwin's teaching to develop the atheistic doctrine of socialism.

2. _____ _____ drew from Darwin's teaching to develop his philosophy of racial superiority and mass extermination of those _"unworthy of life."_

3. The_____ _____ for over forty-five years promoted the atheistic doctrine developed by the author in question #1 and brought many nations to financial, social, and spiritual ruin.

4. Before engaging in "racial cleansing" against the Jews and other races, Adolph Hitler had _____ people killed. These people were eliminated because they were said to be_____.

5. **BONUS:** Euthanasia is practiced in our society today. The common term used to describe this process is_____ _____. This is usually done to the aged or diseased. If this is acceptable, then why couldn't this apply to any and all people wanting to end a miserable existence?

6. "_____ exalts a nation, but _____ is a _____ to any people."

Short Answer and Discussion

1. Besides those mentioned in this lesson, what are other ways that the doctrine of evolution can negatively affect society? _____
_____

2. How do we know that God created all men equally?_____
_____

3. What two parables of Christ did we cite that illustrate what the Christian's attitude toward the less fortunate should be? _____
_____

4. Why might scientific research into evolution be considered wasteful? _____
_____
_____

5. Is all knowledge helpful? What Scriptures speak to this fact?_____
_____
_____

6. Name some ways you can use the material in this book to teach others the truth about the creation versus evolution controversy. What other truth could you teach them?_____
_____
_____

Discussion Question in Preparation for Answering Unbelievers and Critics

Your history teacher, whom you know from previous lectures believes in evolution, makes the following statement in class one day: "Most of today's social ills stem from religion and religious fundamentalism. The world would be a much better place if religion were obliterated." He then invites comments or questions. What do you say? _____
_____
_____
_____
_____

The Darwinist's Hymn
(To the tune of "Give Me That Old Time religion")

Give me that old time evolution, give me that old time evolution, give me that old time evolution. It's good enough for me.

It was good enough for Darwin, it was good enough for Darwin, it was good enough for Darwin, so it's good enough for me.

It's become my religion, it's become my religion, it's become my religion, and it's good enough for me.

Oh, you can't teach any other, you can't teach any other, you can't teach any other. Darwin's good enough for me.

Give me that old time evolution, give me that old time evolution, give me that old time evolution. It's good enough for me.

—Wells, 2005

Appendix A

Should We Use the Term "Neo-Darwinism":
(On the Death of Neo-Darwinism and Twenty-One Proposed Alternative Mechanisms for Macroevolution)

Introduction

The term Darwinism and Darwinian Evolution are technically antiquated terms, because Darwin's theory, based on Lamarckian principles, was replaced in the 1930's by Neo-Darwinism (ND), which was eventually called the "modern synthesis" of evolution (MS) by Julian Huxley in 1942.

Nevertheless, today some evolutionists object to using Neo-Darwinism, neo-Darwinian evolution, or the "modern synthesis" of evolution, and mercilessly criticize creationists for using these textbook terms. Why? Evolutionists are correct in stating that by the 1970's, much of the evolutionary community concluded that Neo-Darwinism (random mutations + natural selection) could *not fully account* for "descent with modification" (i.e., microbe to man evolution). Many new evolutionary mechanisms, therefore, were proposed during this era including "neutral theory" and "punctuated equilibrium," but there was no consensus on which new mechanism should be accepted to replace Neo-Darwinism.

Dr. Larry Moran's attack on Intelligent Design advocates

Thus, there is currently a heated debate on what to call modern humanistic evolution (i.e., the general theory of evolution). For example, University of Toronto atheist professor Dr. Larry Moran (Department of Biochemistry, and rabid anti-creationist and anti-Intelligent Design blogger) said, "I believe that Gould was correct when he pronounced the death of the Modern Synthesis. I agree with him, and with Masa-toshi Nei, that mutation and mutationism were downplayed in the Modern Synthesis. That's one example of why the old-fashioned Modern Synthesis should be abandoned as a description of modern evolutionary theory... The Modern Synthesis has been substantially changed by modern population genetics and Neutral Theory so that it's no longer useful to describe modern evolutionary theory as the 'Modern Synthesis' " (2014).

Elsewhere, Dr. Moran lambasts ID (intelligent design) supporters for using these textbook terms such as "Neo-Darwinism", while calling these people *"IDiots."* He says, "Do you see the strategy? The IDiots are going to claim that their silly misunderstanding of evolution ("Darwinism" or "Neo-Darwinism") is the "fact" of evolution that's being taught in the schools. Thus, it's not their fault that their understanding of evolution is wrong—blame it on the evolutionary biologists. Yeah, that'll work! :-) The alternative is to admit that the IDiots are, well... idiots. Don't hold your breath waiting for that to happen" (Moran, 2011). Again, Moran skewers *Discovery Institute* Fellow and attorney, Casey Luskin by saying, "Casey Luskin knows d___ well that the 'Darwinian Theory' he talks about is not equivalent to modern evolutionary theory and that's what the 'mainstream technical literature' says. He knows d___ well that his flock of followers will interpret his statements to mean that legitimate scientists are skeptical about evolution" (Moran, 2015b).

Further, Moran says, "The reason this is so upsetting is that the IDiots know full well that the complexity of life is due to far more than just mutation + selection. They also know that evolutionary biologists have been 'examining' Darwinian theory for a century and have decided that it is not sufficient to account for evolution. The obvious question is: are IDiots stupid or liars, or both?" (Moran, 2015a). Well, thanks for admitting what thousands of scientists refuse to admit, Dr. Moran; namely, that *Neo-Darwinism cannot account for macroevolution,* even though it's taught to our kids in textbooks in all fifty states.

To corroborate Dr. Moran's statement, Dr. Lynn Margulis (Darwin-Wallace Medal winner), stated, "At that meeting Ayala agreed with me when I stated that this doctrinaire Neo-Darwinism is dead. He was a practitioner of Neo-Darwinism, but advances in molecular genetics, evolution, ecology, biochemistry, and other news had led him to agree that Neo-Darwinism is dead" (Margulis, 2010, p. 285).

Responding to Dr. Larry Moran

A possible substitute for the ND or MS nomenclature problem may be the terms the *"extended synthesis of evolution"* or the *"extended modern synthesis of evolution,"* but there is no consensus on using those terms, so most people still use ND or MS.

After all the alternatives to Neo-Darwinism had been considered, Stephen J. Gould concluded, "I recognize that we know no mechanism for the origin of such organismal features other than conventional natural selection at the organismic level" (2002). Dr. Muller and Newman also affirm, "the neo-Darwinian paradigm still represents the central explanatory framework of evolution" (2003).

Oddly enough, on Larry Moran's own web page he stated, "Some scientists continue to refer to modern evolutionary theory as Neo-Darwinian. In some cases these scientists do not understand that the field has changed, but in other cases they are referring to what I have called the Modern Synthesis, only they have retained an old name from the early 1900s" (Moran, 2009). Can you see the double standard here? It's okay when Dr. Moran's colleagues use the terms ND and MS, just not the creationists.

Earlier, Dr. Moran on the ultra-evolutionary "TalkOrigins" website gave more credence to the use of the terms ND and the MS in present-day literature as follows: "more recently the classic Neo-Darwinian view has been replaced by a new concept which includes several other mechanisms in addition to natural selection. Current ideas on evolution are usually referred to as the Modern Synthesis... Some scientists continue to refer to modern thought in evolution as Neo-Darwinian. In some cases these scientists do not understand that the field has changed, but in other cases they are referring to what I have called the Modern Synthesis, only they have retained the old name" (Moran, 1993).

The Oxford Dictionary of Biology also gives support for the use of the term ND and defines *"Neo-Darwinism"* thusly: "The current theory of the process of evolution, formulated between 1920 and 1950, that combines evidence from classical genetics with the Darwinian theory of evolution by natural selection...it makes use of modern knowledge of genes and chromosomes to explain the source of genetic variation upon which selection works. This source was unexplained by traditional Darwinism" (Martin and Hine, 2008).

Here is one summary of the naming controversy and use of the term Neo-Darwinism: "Following the development, from about 1937 to 1950, of the modern evolutionary synthesis, now generally referred to as the synthetic view of evolution or the modern synthesis, the term neo-Darwinian is often used to refer to contemporary evolutionary theory. However, such usage has been described by some as incorrect; with Ernst Mayr writing in 1984 that 'the term Neo-Darwinism for the synthetic theory is wrong, because the term Neo-Darwinism was coined by Romanes in 1895 as a designation of Weismann's theory.' Despite such objections, publications such as *Encyclopedia Britannica* use this term to refer to current evolutionary theory. This term is also used in the scientific literature, with the academic publisher Blackwell Publishing referring to "Neo-Darwinism as practiced today," (2015) and some figures in the study of evolution like Richard Dawkins (2010) and Stephen Jay Gould (1984) using the term in their writings and lectures" (Wikipedia, 2015, "Neo-Darwinism").

Also, from the *New World Encyclopedia* (2010): "It is felt by some that the term 'Darwinism' is sometimes used by creationists as a somewhat derogatory term for 'evolutionary biology,' in that casting of evolution as an 'ism'—a doctrine or belief—strengthens calls for 'equal time' for other beliefs, such as creationism or intelligent design. However, top evolutionary scientists, such as Gould and Mayr, have used the term repeatedly, without any derogatory connotations."

Even Richard Dawkins, in the Dawkins-Lennox debate, used the terms Darwinism and Darwinian *twelve times* to refer to the modern theory of naturalistic evolution (Dawkins and Lennox, 2007).

Jerry Coyne used the term Darwinism multiple times in his article, "Why Evolution Is True" (2009).

Referring to modern naturalistic macroevolution, atheist Dr. Michael Ruse said, "And so, we now have this synthetic theory. The synthesis of Darwin's theory and Mendel's theory or as it known in England as Neo-Darwinism, which is a full blooded causal theory of Darwinian selection, plus modern molecular genetics... But the point is it's all done within the Darwinian paradigm" (2009).

Casey Luskin's response

Casey Luskin (2015) of the *Discovery Institute* (at the highly recommended website [*"Evolution: News and Views"*]) has pointed out bona fide evolutionists who continue to use the term Neo-Darwinism, despite Dr. Moran's railing, as follows:

Douglas Futuyma's 2005 textbook Evolution defines *"Neo-Darwinism"* as "the modern belief that natural selection, acting on randomly generated genetic variation, is a major, but not the sole, cause of evolution" (Futuyma, 2005).

Strickberger's textbook Evolution equates *"Neo-Darwinism"* with the *"modern synthesis,"* defining it as "a change in the frequencies of genes introduced by mutation, with natural selection considered as the most important, although not the only, cause for such changes" (Strickberger, 2000).

A letter by scientists published in one of the world's top scientific journals, *Nature*, states: "The two central elements of neo-darwinian evolution are small random variations and natural selection" (Newman *et al.*, 1985).

A paper by two scientists in the journal *Science*, a top-tiered scientific journal, notes: "According to neo-Darwinian theory, random mutation produces genetic differences among organisms whereas natural selection tends to increase the frequency of advantageous alleles" (Lenski and Mittler, 1993).

Which proposed evolutionary mechanism will be chosen to extend or replace Neo-Darwinism?

HERE'S THE POINT: The reason for needing a name change in the first place revolves around an even more vehement debate currently raging in the evolutionary community. Namely, *which evolutionary mechanism(s) should be used to extend or supplant the "dead" ND and MS?* This would also be a good question to ask your macroevolution believing friends. Ask them, "Which of the following twenty-one proposed mechanism(s), do you feel is the most tenable to 'extend' or replace the modern synthesis of evolution (Neo-Darwinism)? Evo-devo? Genetic flow? Genetic hitchhiking? Genetic drift? Biased Mutation? Epigenetic Inheritance? Niche Construction? Self-organization (Kauffman et al.)? Whole Genome Doubling (Shapiro)? Orthogenesis (Directed Evolution) (Lima-de-Faria)? Saltationism (Balon, Norrstrom)? Process Structuralism? Symbiotic reorganization (Margulis et al.)? Punctuated equilibrium? Natural genetic engineering (Shapiro)? Neo-Lamarckism (Jablonka, Pigliucci)? Facilitated variation (Gerhart and Kirschner)? The Gaia hypothesis? Directed mutation (adaptive evolution) (Cairns)? Teleonomic Selection (Corning)? Nearly neutral theory? etc. etc."

They will likely be confused and may try to divert the discussion from the undeniable problems with current ND theory. Don't be shaken. Keep pressing the issue that ND is dead, and that there is no consensus among the scientific community on an alternative mechanism(s). Amidst your discussions, you may also want to read *"Darwinian Debating Devices"* and *"Frequently Raised but Weak Arguments Against Intelligent Design"* at the Uncommon Descent website (2015).

Conclusion

Until there is a consensus among the scientific community as to what new names or descriptors we should use to replace ND and the MS, we should continue to use the textbook terms *"Neo-Darwinism," "neo-Darwinian* evolution," the "modern synthesis" of evolution, and the "modern evolutionary synthesis" to refer to currently accepted naturalistic biological evolutionary theory.

Acknowledgment

Many thanks to Amanda Smelser for reviewing this article and providing helpful feedback.

REFERENCES

Ager, Derek V. 1993a. *The Nature of the Stratigraphical Record.* Hoboken, NJ:Wiley Publishing. 166 pp.

_____. 1993b. *The New Catastrophism: The Importance of the Rare Event in Geological History.* Cambridge:Cambridge University Press. 231 pp.

Behie, Alison M., and Marc F. Oxenham. 2015. *Taxonomic Tapestries: The Threads of Evolutionary, Behavioural and Conservation Research.* p. 157, Anu Press. 415 pp.

Blackwell Publishing. 2015. "Definition of Neo-Darwinism." As quoted in *Evolution, Third Edition,* by Mark Ridley. Accessed online on 3/16/15 at: http://www.blackwellpublishing.com/ridley/a-z/Neo-Darwinism.asp

Coyne, Jerry. 2009. "Why Evolution is True," *Forbes Magazine.* Accessed online on 3/21/15 at: http://www.forbes.com/2009/02/12/evolution-creation-proof-opinions-darwin_0212_jerry_coyne.html

Dawkins, Richard. 2010. "Lecture on Neo-Darwinism." Presented on the Galapagos Islands. Accessed online on 3/16/15 at: http://old.richarddawkins.net/videos/1345-lecture-on-neo-darwinism

Dawkins, Richard, and John Lennox. 2007. *The God Delusion Debate.* Birmingham, AL. October, 3. Accessed on 3/18/15 at: http://fixed-point.org/index.php/video/35-full-length/164-the-dawkins-lennox-debate TRANSCRIPT available on 3/18/15 at: http://www.protorah.com/god-delusion-debate-dawkins-lennox-transcript/

Dott, R.H., Jr. 1998. "What is unique about geological reasoning?" *GSA Today.* 8(10):15-18.

Futuyma, Douglas J. 2005. *Evolution,* p 550. Sinaur Publishers:Sunderland, MA.

Gould, S.J. 1965. "Is Uniformitarianism Necessary?" *American Journal of Science* 263:223-228.

_____. 1984. "Toward the Vindication of Punctuational Change." In: W.A. Berggren and J.A. Van Couvering, *Catastrophes and Earth History: The New Uniformitarianism.* Princeton, NJ:Princeton University Press. 478 pp.

_____. 1984. "Challenges to Neo-Darwinism and Their Meaning for a Revised View of Human Consciousness." *The Tanner Lectures on Human Values,* Delivered at Clare Hall, Cambridge University, April 30 and May 1. Accessed online at: http://tannerlectures.utah.edu/_documents/a-to-z/g/gould85.pdf

_____. 2002. *The Structure of Evolutionary Theory.* Harvard University Press (Belknap Press): Cambridge, Massachusetts. 1464 pp., p. 710.

Gretener, Peter, E. 1984. "Reflections on the 'Rare Event' and Related Concepts in Geology." In: W.A. Berggren and J.A. Van Couvering, *Catastrophes and Earth History: The New Uniformitarianism.* Princeton, NJ:Princeton University Press. 478 pp.

Huxley, Julian. 1942. *Evolution: the modern synthesis.*

Lenski, R.E., and J.E. Mittler. 1993. "The directed mutation controversy and Neo-Darwinism," *Science*, 259:188-194.

Luskin, Casey. 2015. "Answering Objections to the Dissent from Darwinism List," *Evolution: News and Views Website,* Accessed on 3/21/15 at: http://www.evolutionnews.org/2015/03/answering_objec_2094331.html

Martin, Elizabeth, and Robert Hine. 2008. *Oxford: A Dictionary of Biology*, 6th edition, Article: "Neo-Darwinism," Accessed online on 3/21/15 at: http://oxfordindex.oup.com/view/10.1093/oi/authority.20110803100228213?rskey=SdRY2T&result=0&q=Neo-Darwinism

Moran, Laurence A. 1993. "The Modern Synthesis of Genetics and Evolution." *TalkOrigins Website*. Accessed on 3/16/15 at: http://www.talkorigins.org/faqs/modern-synthesis.html

_____. 2009. "The Modern Synthesis" on *Sandwalk: Strolling with a skeptical biochemist*. Accessed online on 3/17/15 at: http://sandwalk.blogspot.com/2009/02/modern-synthesis.html

_____. 2009. "Casey Luskin Is Confused (Again)" on *Sandwalk: Strolling with a skeptical biochemist*. Accessed online on 4/13/15 at: http://sandwalk.blogspot.com/2011/04/casey-luskin-is-confused-again.html

_____. 2014. "Rethinking Evolutionary Theory." Thurs. Oct. 9. on *Sandwalk: Strolling with a skeptical biochemist*. Accessed online on 3/19/15 at: http://sandwalk.blogspot.com/2014/10/rethinking-evolutionary-theory.html#more

_____. 2015a. "Apparently it really is impossible to teach Intelligent Design Creationists about evolutionary theory." Tues. March 17. on *Sandwalk: Strolling with a skeptical biochemist*. Accessed online on 4/13/15 at: http://sandwalk.blogspot.com/2015/03/apparently-it-really-is-impossible-to.html

_____. 2015b. "Is it impossible to educate Intelligent Design Creationists on evolutionary theory?" March 12. On *Sandwalk: Strolling with a skeptical biochemist*. Thurs. Accessed online on 4/13/15 at: http://sandwalk.blogspot.com/2015/03/is-it-impossible-to-educate-intelligent.html

Muller, Gerd B., and Stuart A. Newman. (2003), "Origination of Organismal Form: The Forgotten Cause in Evolutionary Theory," In: Ed. G.B. Muller and S.A. Newman, *Origination of Organismal Form: Beyond the Gene in Developmental and Evolutionary Biology,* pp. 3-10, MIT Press: Cambridge, MA.

New World Encyclopedia. 2010. "Darwinism," Accessed online on 3/17/15 at: http://www.newworldencyclopedia.org/entry/Darwinism

Newman, C. M., J. E. Cohen, and C. Kipnis. (1985), "Neo-darwinian evolution implies punctuated equilibria," *Nature*, 315:400-401.

Ruse, Michael. 2009. "Is Darwinism past its 'Sell By' date?" a lecture presented as part of the Darwin Distinguished Lecture Series, Arizona State University, Office of the President, College of Liberal Arts and Sciences, School of Life Sciences, and the Center for Biology and Society. February 6. Accessed online on 3/21/15 at: http://darwin.asu.edu/media/transcripts/04_ruse.php

Shea, James H. 1983. "Creationism, Uniformitarianism, Geology and Science." *J. Geologic. Edu.* 31:105-110.

Strickberger, Monroe W. 2000. *Evolution*, p. 649 3d Ed., Jones and Bartlett Learning:Burlington, MA, 722 pp.

Uncommon Descent: Serving the Intelligent Design Community. 2015. "Darwinian Debating Devices" and "Frequently Raised but Weak Arguments Against Intelligent Design" accessed on 3/17/15 at: http://www.uncommondescent.com/category/ddd/ AND http://www.uncommondescent.com/faq/

Appendix B1

The Law of Noncontradiction Argues for the God of the Bible and against Atheism, Hinduism and Buddhism

There is an argument for God that I would like for you to consider. This argument says that without the Christian worldview (which is simply the comprehensive belief system, as taught in the Bible), nothing in the world could make rational sense. This is based on three sets of natural laws supported in the Scripture:

1. **Laws of Logic** (e.g., the law of noncontradiction (LNC))

2. **Uniformity of Nature** (Laws of nature, the preconditions of science)

3. **Laws of morality**

This is known as the presuppositional approach to apologetics. No other worldview than that of the Bible, including atheism, Buddhism or Hinduism, can account for these three laws. In this article, we will only focus on a subgroup of Law #1, which is the law of noncontradiction (LNC). The LNC states that you cannot have A and not-A existing at the same time in the same relationship. Put into practice, the LNC means you cannot have an argument or statement that is self-contradictory (which includes lies). According to the LNC, I cannot affirm that my only dog is dead and simultaneously affirm that my only dog is alive.

1. Atheism cannot account for the Law of Noncontradiction

Atheism is typically a belief firmly rooted in materialism. That is, atheism teaches that the material universe is all that exists. However, laws of logic and the LNC are non-material abstract laws that make no sense in an atheistic material universe. According to atheists, our universe and everything in it is a product of purposeless, non-guided, random chance accidents. There are numerous quotes where atheists affirm this.

"Such a hypothesis leads to a view of creation in which the entire universe is *an accident.* In Tyron's words, 'Our universe is simply one of those things which happen from time to time.'"—Dr. James Trefil, "The Accidental Universe," *Science Digest,* vol. 92 (June 1984), pp. 53-55, 100-101.

Therefore, in a random chance, happenstance, accidental, unguided material universe, abstract, absolute physical laws would not exist. There is no reason that physical laws should exist and be consistently applied in all parts of an accidental universe. Nevertheless, our universe is extremely fine tuned (anthropic principle) by dozens of known unchangeable laws such as the *strong and weak laws of gravity, Boyle's law, the four laws of thermodynamics, laws of electromagnetism, laws of photonics, laws of quantum mechanics, laws of electromagnetic radiation, laws of gravitation,* and *relativity,* etc. Laws, including the LNC, must come from a lawgiver, a mind, which must have preceded the material universe. God.

2. Hinduism cannot account for the Law of Noncontradiction

Laws of logic and the LNC demand a rational reasoned response to a question, which assumes we can formulate reasonable answers by use of our senses. Hinduism, on the other hand, teaches the doctrine of *Maya,* which states that everything observed within normal human consciousness are simply illusions. Thus, we are trapped in a Hindu world of illusions where right is wrong, black is white, up is down, and the LNC is negated. Consider the following Hindu quotes:

"The world in which we live is also not very different from the hall of illusions we read about in the Mahabharata. We also live here enveloped by illusion, in a state of ignorance about ourselves, whereby we fail to discriminate between truth and falsehood, What is Truth? We all suffer from the grand illusion that what we know and experience through our senses is the truth and that we are capable of knowing the facts of our existence with the help of our minds and senses, where as the truth is we cannot discern reality with our limited consciousness. We cannot answer the question about truth truthfully, because we do not know the answer. We may give an answer, some answer, but that answer would not be correct." Accessed on 2-28-15 at http://www.hinduwebsite.com/maya.asp

Contrast this with Jesus' teaching, which says we can know the truth: "Then Jesus said to those Jews who believed Him, 'If you abide in my word, you are my disciples indeed. And you shall know the truth, and the truth shall make you free" (John 8:31-32).

3. Buddhism and Taoism cannot account for the Law of Noncontradiction

Integral to the teaching of modern Buddhism and Taoism is the *doctrine of contradiction*. Both religions claim that contradictions are acceptable and should be incorporated into our reasoning, which denies the law of noncontradiction.

Dr. Greg Bahnsen (historic apologist who was a known master of these arguments) used to say that if anyone denied the law of noncontradiction, you could tell them "oh, so you don't deny it." Then they would respond, "No, I do deny it." Then you would respond, "Yes, but if you deny it, then you also don't deny it."

Here are some descriptions of Buddhism and Taoism denying the LNC:

"Buddhist logicians sometimes added a fifth possibility: none of these. (Both positions were called the *catushkoti*.) The Jains went even further and advocated the possibility of contradictory values of the kind: true (only) and both true and false. (Smart, 1964, has a discussion of the above issues.) Contradictory utterances are a commonplace in Taoism... When Buddhism and Taoism fused to form Chan (or Zen, to give it its Japanese name), a philosophy arose in which

contradiction plays a central role. The very process for reaching enlightenment (Prajna) is a process, according to Suzuki (1969, p. 55), which is at once above and in the process of reasoning. This is a contradiction, formally considered, but in truth, this contradiction is itself made possible because of Prajna." Accessed on 2-28-15 at http://stanford.library.usyd.edu.au/archives/fall2008/entries/dialetheism

"As Buddhism evolved over the centuries, many different authors from varying cultures set forth their own ideas in the name of the Buddha. As a result, Buddhism developed inherent contradictions. When this was realized, Buddhism embraced these contradictions as a badge of honor. Thus the making of self-contradictory statements has become one of the pronounced features of Zen and other esoteric forms of Buddhism." Accessed on 2-28-15 at http://www.faithdefenders.com/articles/worldreligions/Buddhism_Unmasked.html

4. The only rational, reasonable, noncontradictory alternative is the God of the Bible

Both Hinduism and Buddhism construct a universe that denies simple laws of logic leading to a confusing, nonsensical, irrational, and contradictory existence. The God of the Bible, on the other hand, demands that humans use reason and rationality via the LNC to understand His word and understand truth. In this respect, R.C. Sproul said:

"Logic is like a policeman that God has put in the brain of human beings, to blow the whistle; to recognize the lie. The whistle blows, and things don't compute. Just like your computer goes whacky when you ask it to be irrational. So, God has built into the human mind a function of rationality that is a test of coherency. A test of rationality. And at the very heart of the Christian affirmation, is that. Though the content that we get in the Bible goes far beyond what we can learn, through rational speculation. It's based on Divine revelation. That Divine revelation does not come to us packaged in absurdity. The Word of God is not... irrational. It is addressed to creatures who have been given minds that operate according to these principles."

5. Bible passages that affirm the Law of Noncontradiction

A. God cannot lie nor deny Himself

Hebrews 6:17-18

"So when God desired to show more convincingly to the heirs of the promise the unchangeable character of his purpose, he guaranteed it with an oath, so that by two unchangeable things, in which it is impossible for God to lie, we who have fled for refuge might have strong encouragement to hold fast to the hope set before us."

1 Corinthians 14:33

"For God is not a God of confusion but of peace."

1 Timothy 2:13

"If we are faithless, He remains faithful; He cannot deny Himself."

Titus 1:2

"In hope of eternal life which God, who cannot lie, promised before time began."

2 Corinthians 1:17-20

"Therefore, when I was planning this, did I do it lightly? Or the things I plan, do I plan according to the flesh, that with me there should be Yes, Yes, and No, No? But as God is faithful, our word to you was not Yes and No. For the Son of God, Jesus Christ, who was preached among you by us—by me, Silvanus, and Timothy—was not Yes and No, but in Him was Yes. For all the promises of God in Him are Yes, and in Him Amen, to the glory of God through us."

B. God made us in His image and expects us to think like Him and imitate Him (Gen. 1:26; I Pet. 1:16; Eph. 5:1)

C. Therefore, God expects us to use our intellect and our built in laws of logic to understand Him by understanding His word

Ephesians 5:17

"Therefore do not be unwise, but understand what the will of the Lord is."

Romans 12:1

"I beseech you therefore, brethren, by the mercies of God, that you present your bodies a living sacrifice, holy, acceptable to God, which is your reasonable service."

Reasonable is from *logikos*, which is logical or rational. In order for us to even understand God's word we must use our reasoning and intellect, which implies noncontradictions. If we *assume* contradictions, then we cannot rationally/logically make sense of any statement.

Does the Bible contain mysteries? Certainly. But, it doesn't imply that logical contradictions are acceptable, as does Buddhism, which is why this is a good argument for the God of the Bible and against atheism and the eastern religions. The whole of Christ's teaching and the New Testament is a testimony to the fact that we are to use logical reasoning to believe and obey Him (Heb. 5:9).

Appendix B2

Recommended Creation-Evolution / Origins Materials

(Books with asterisks [*] are highly recommended for all libraries. Two asterisks [**] are essentials).

A. **Free Monthly Magazines**
1. *Acts and Facts* by ICR (Institute for Creation Research). Sign up at www.icr.org
2. *Answers in Genesis Newsletter.* Sign up at www.answersingenesis.org

B. **Books** (Find these at Amazon.com, Answersingenesis.org, etc.)
1. *** Refuting Evolution**, by Jonathan Sarfati, Ph.D.
2. *** Refuting Evolution Part 2**, by Jonathan Sarfati, Ph.D.
   - Both of Dr. Sarfati's books are succinct and very readable, even for those at the junior high level.
3. **** The New Answers Book: Parts I, II, III and IV**
   - No library should be without this essential series. Answers over 100 of the most asked Creation/Evolution and origins questions by numerous credentialed experts, edited by Ken Ham.
4. *** Creation: Facts of Life**, by Gary Parker, Ph.D.
   - The 2006 version is revised and updated.
5. *Evolution: A Theory in Crisis*, by Michael Denton, Ph.D.
   - This book is slightly heavier reading than the others in this list, but is the classic 1980's work by a renowned Australian biochemist. *Evolution: A Theory in Crisis* has influenced many scientists (one of whom was

Darwin's Black Box author, Michael Behe) to begin questioning the validity of naturalistic evolution. (NOTE: Old-earth creationist)
6. *Not by Chance!*, by Lee Spetner, Ph.D.
   - Dr. Spetner is a Massachusetts Institute of Technology (MIT) graduate, Israeli physicist who was employed by John's Hopkins University to study guided-missile systems. (NOTE: Old-earth creationist)
7. **** Icons of Evolution**, by Jonathan Wells, Ph.D.
   - An excellent overview of the fallacies of some of the more commonly used historical textbook arguments for Darwinian evolution.
8. *** The Scientific Case for Creation**, by B. Thompson, Ph.D.
   - This book is available for purchase through Apologetics Press or can be downloaded for *FREE* in PDF format from www.ApologeticsPress.org (NOTE: Check out AP's entire FREE pdf book series).
9. *** The Modern Creation Trilogy, Volume 2: Science and Creation**, by Henry M. Morris, Ph.D. and John D. Morris, Ph.D. (My personal favorite. You won't be disappointed, JBG).
10. *The Altenberg 16: An Exposé of the Evolution Industry*. Suzan Mazur. 2010. North Atlantic Books: Berkeley, CA. 376 pp. [Exposes the schism between "old" neo-darwinists and the modern anti-ND macroevolutionists]
11. *What Darwin Got Wrong*. Dr. Jerry Fodor, with Dr. Massimo Piattelli-Palmarini. 2010. Farrar, Straus and Giroux Publishers: New York, New York. 320 pp. (Sold out and reprinted by Picador, 2011). [Exposes the schism

between "old" neo-darwinists and the modern anti-ND macroevolutionists.]

C. **Websites** (Notes: [1] Websites classified as "old-earth creation" {OEC} or "young earth creation" {YEC}, [2] Be aware of Calvinism and premillennialism).

1. ***Answers in Genesis***, www.answersingenesis.org Evangelically-based apologetics organization founded by Ken Ham. (YEC)
2. ***The Institute for Creation Research***, www.icr.org Evangelically-based apologetics organization founded by Dr. Henry Morris. (YEC)
3. ***Apologetics Press***, www.ApologeticsPress.com Evangelically-based apologetics organization. (YEC)
4. ***Creation Wiki***—Wikipedia-type website for creation material www.creationwiki.org (YEC)
5. ***Creation Ministries International-Australian-based***. Excellent scientific articles. www.creation.com (YEC)
6. ***Evolution News and Views***—A premier Intelligent Design website. Posts the latest information on the creation/evolution controversy. www.evolutionnews.org (Mainly OEC)
7. ***Uncommon Descent***—Another premier ID website originally created by Dr. William Dembski. NOTE: Most but not all individuals posting on this page are "old-earth creationists." www.uncommondescent.com (Mainly OEC)
6. ***True Origins.*** www.trueorigin.org Started in response to the atheistic "Talk Origins" organization. (Mainly OEC)
7. ***Discovery Institute.*** Intelligent Design organization based in Seattle Washington, featured in Ben Stein's movie, "*Expelled.*" www.discovery.org Not "Biblical" *per se*, but is considered a "big tent" for anyone who opposes Neo-Darwinism. (Mainly OEC)
8. ***Dissent from Darwin.*** Petition of 200+ doctorates who oppose Neo-Darwinism. www.dissentfromdarwin.org (OEC and YEC signers)
9. ***Physicians and Surgeons for Scientific Integrity.*** A petition of 868 physicians and surgeons who oppose Neo-Darwinism www.pssiinternational.com (OEC and YEC signers)

Appendix C1

Dangers of Rejecting the *Unambiguous* Twenty-Four-Hour Days of Creation

Introduction

If God cannot be understood in the straightforward, unambiguous narrative of Genesis one, why should we believe His straightforward, unambiguous teaching anywhere else in Scripture? Teaching young people that "believers have misunderstood the word *day* in Genesis one for the last 3,500 years, but now, based on modern science, we know the word *day* cannot actually mean *day* as we thought" opens the door of subjugating Scripture to popular science. The man in the white lab coat, and not God's word, is now the ultimate standard (Ps. 11:3; 119:89).

How to respond to the charge, *"the literal days of Genesis one are unimportant"*

When one asserts that, *"the length of the days in* Genesis *one are unimportant"* and we would do better to focus on *"important doctrinal matters"* such as the virgin birth, the resurrection, sin, salvation, eternal punishment, and eternal life, ask the following question: "How do you know there was a virgin birth, a resurrection, and that there is sin, Satan, salvation, eternal punishment, and eternal life?" The only way we can know any of these facts is because the Bible tells us in clear unambiguous language (Eph. 5:17)—always based on the *context*. Arguing that the days of Genesis one are unimportant inadvertently undermines the validity of all the doctrinal issues mentioned above, by requiring their veracity to hinge on verification/validation by popular science.

If we can't understand the unambiguous word *day* in

Genesis one, can we understand any unambiguous language in the Bible?

If the word *day* in Genesis one can only be understood based on *popular science*, why not apply the same principle to the rest of Scripture? (e.g., the virgin birth, the resurrection, sin, salvation, Satan, eternal punishment, and eternal life). If we can't *know* that the use of the word *day* in Genesis one is a literal day, how can we know Mary was really a *virgin*? How can we know Jesus really died on the cross? How can we know that the portions of three days Jesus was in the tomb were literal twenty-four-hour days and that He really rose from the dead (1 Cor. 15:29-32)? How can we know *eternal* really means *eternal*, that sin is really rebellion against God, that Satan is really a demonic force, or that eternal punishment truly exists for the unsaved?

At least since the rise of popular liberal theology in eighteenth-century Western Europe, skeptics have effectively chipped away at the faith of believers by supplanting Scripture with scientific hypotheses. Yet, when believers reject the unambiguous teaching in the very first chapter of the Bible with regard to the word *day*, they are unintentionally playing right into the critics' hand. The book of Genesis should be the beginning of our faith, not the beginning of unbelief.

Why divisive controversy over the length of "day" only in Genesis one?

Whenever the word *day* in the Bible is used in reference to a time other than a twenty-four-hour period (e.g., a protracted period of time, daylight or waking hours, etc.) this meaning is always *contextually* manifest and never in dispute. Although the Hebrew words for *day* and *days* appear over 2,300 times in the Old Testament, divisive controversy only exists over their meaning in Genesis one. Why? Here is one explanation. For the last 200 years, believers have been pressured into melding the supposed scientific age of the earth (currently ca. 4.5 billion years) with the Bible; and Genesis one is the only place believed that the eons of time can be inserted.

Evidence that the days of Genesis one are literal twenty-four-hour days

1. In the Old Testament, outside of Genesis one, when the word *day* (*yom*) is modified by a number (over 400 times), it always refers to a normal twenty-four-hour day (exceptions [Isa. 9:14; 10:17; 47:9], may refer to literal twenty-four-days of divine destruction, judgment or extremely short periods, but not ages).

2. In the Old Testament, outside of Genesis one, when the word *day* is used with the words "evening" *and* "morning" together (thirty-eight times), it always refers to a normal twenty-four-hour day.

3. In the Old Testament, outside of Genesis one, when the word *day* is used with the words "evening" *or* "morning" individually (twenty-three times), it always refers to a normal twenty-four-hour day.

4. In the Old Testament, outside of Genesis one, when the word *day* is used with the word "night" (fifty-two times), it always refers to a normal twenty-four-hour day.

Based on points 1-4 above, consider the following combinations of words used with the word *day* in Genesis one:

Verse 5—Night, evening, morning, first day

Verse 8—Evening, morning, second day

Verse 13—Evening, morning, third day

Verse 14—Day, night

Verse 16—Day, night

Verse 18—Day, night

Verse 19—Evening, morning, fourth day

Verse 23—Evening, morning, fifth day

Verse 31—Evening, morning, sixth day

Could God be hinting at something? What could God have said to make the meaning any

clearer? The apostle Paul, on one occasion, warned of exacting more of God than is necessary (Rom. 10:6-8).

Further, the days of creation are defined in unambiguous language by the Lord in Exodus 20:11 and 31:17. It is interesting that the only occasion that Jehovah penned *His* law for *His* people *by His own* hand was on tablets of stone in Exodus twenty. If we cannot believe God when He unambiguously pens *His* own words by *His* own hand into tablets of stone, when can we believe Him? Specifically, He wrote, "Remember the Sabbath day, to keep it holy. *Six days* you shall labor and do all your work, but the *seventh day* is the Sabbath of the Lord your God. In it you shall do no work: you, nor your son, nor your daughter, nor your male servant, nor your female servant, nor your cattle, nor your stranger who is within your gates. For in *six days* the Lord made the heavens and the earth, the sea, and all that is in them, and rested *the seventh day.* Therefore the Lord blessed *the Sabbath day* and hallowed it" (Exod. 20:8-11).

How can we know when the word *day* is a twenty-four-hour day?

If the days in Genesis one aren't literal twenty-four-hour days, and we are told we need corroborating scientific evidence to conclude they were twenty-four-hour days, when can we ever be sure that *day* means a literal day anywhere else in Scripture? For example, over what period of time did the children of Israel march around Jericho? Seven twenty-four-hour days. But, how can we know those were real days? Maybe each day really represented an hour... or a month... or a year. What about Jonah's three days in the fish or the Christ spending portions of three days in the grave? Are these literal historical events or metaphor? Must we demand corroborating scientific evidence to believe that they occurred?

Interpreting the words day (*yom*) and days (*yamim*) as twenty-four-hour periods of time in Genesis one follows clear hermeneutical rules every Christian uses to decipher God's word (or any other written word, for that matter). Reinterpreting the twenty-four-hour days in Genesis one to accommodate popular science (if consistently applied to the rest of Bible) so damages a simple understanding of Scripture that nothing in the Bible can be believed unless the man in the white lab coat gives the nod. This is the reasoning of a "natural man" who can no longer be taught the "the things of the Spirit of God" (1 Cor. 2:14).

Do you reinterpret the unambiguous word *day* in Genesis one?

The six twenty-four-hour days of creation were the basis for the six-day Hebrew work week as well as the seven-day week, observed around the world today. The seventh twenty-four-hour day, when the Lord rested from His labors, was also the basis for the Hebrew Sabbath day of rest. Yet, no one questions the length of that day.

Maybe Genesis one is the only passage where you reinterpret God's unambiguous word to agree with secular science. If so, then consider the following two thoughts:

1. To reject unambiguous *contextual* language in Genesis one and accept it in the rest of Scripture is neither consistent nor rightly handling the Word of truth (2 Tim. 2:15).

2. If you inconsistently reject the *contextually* unambiguous language in Genesis one, yet accept *contextually* unambiguous language elsewhere in Scripture, those whom you influence may apply your rule consistently to the rest of Scripture, and subjugate God's word to secular philosophy and science (Judges 2:10). The man in the white lab coat, and not God's inspired word, becomes the ultimate standard. "If the foundations are destroyed, what can the righteous do?" (Ps. 11:3).

Acknowledgments

Thanks to Jeff Archer, Edwin Crozier, Jerry Falk, Phil Martin, Jeff May, Terry Mortenson, Bob Myhan, Joe Price, Dennis Scroggins, and Amanda Smelser for reviewing this article and providing helpful feedback.

Appendix C2

Leveraging "Intelligent Design" with the Bible

(An Appeal to Plant Seeds of Biblical Faith in Skeptics)

Introduction

Some argue that, when demonstrating God's existence to skeptics, we should shun the Bible and use science alone (lest we drive unbelievers away). Although Scripture certainly uses scientific "natural revelation" to corroborate deity (Psa. 19:1-6; Job 12:7-10; Jer 5:23-25; Acts 14:16,17; Rom 1:19,20), should the gospel be imparted *only after* an unbeliever is convinced that God exists? How many conversations are required before we confess the Messiah? I will outline, here, why I think using science, *alone*, is a mistaken apologetic.

The Intelligent Design Movement

Recent decades have seen an explosion in the so-called "Intelligent Design Movement" (IDM). The IDM was launched by Dr. Phillip Johnson from U.C. Berkley, using science (ID) to defend the Creator, without resorting to traditional "Biblical creationism." Capable individuals such as Michael Denton, Stephen C. Meyer, William Dembski, Douglas Axe, Michael Behe, *et al.* have broken new ground, engaging scientific academia in a dialogue regarding ID —where believers had once been stone-walled. ID supporters have experienced intense, often unfair scrutiny—some losing their jobs.

Phillip Johnson rightly presents IDM as a "big umbrella" under which all religious (or irreligious) persuasions can unite. Is finding common ground with others to oppose naturalistic evolution commendable? Yes. Remember, however, that converting the whole world to ID without winning their hearts to Christ, ultimately loses the war for their souls (John 3:15,18; 5:24; 6:40; 14:6; 15:5-6; Acts 4:12; 17:30; 1 Thess. 1:8-9).

Ancient, unsaved, Intelligent Design advocates

IDM is composed of Muslims, Jews, and agnostics, in addition to Christians and other purported Bible-based faiths. Some of these deny the Christ, the inspiration of Scriptures, or even a personal God. Anyone who supports ID is certainly off to a good start, but belief in ID doesn't guarantee that they have a sufficient understanding of Jesus Christ.

Further, remember that countless pagan cultures in the Old Testament would have heartily supported ID. Consider the Canaanites, Hittites, Amorites, Perizzities, Hivites, Jebusites, Ammonites, Moabites, Edomites, Gergashites, Sidonians, Philistines, Amalekites, Egyptians, Babylonians, Assyrians, Medes and Persians who, although polytheists, would have vociferously opposed the Richard Dawkins of their day. Yet (with notable exceptions), these cultures, were lost in sin, when God punished them (Gen. 15:16; Deut. 7:1-2; Matt. 23:32; 1 Thess. 2:15-16).

An appeal to plant seeds of biblical faith in unbelievers

Instead of excluding Scripture and relying on science alone, ID can be a tool for pointing people *back* to Jehovah and *back* to His inspired word (Rom. 10:17). Using natural revelation—ID/science—(Ps. 19:1-6) to reflect divinity is not mutually-exclusive with planting the seeds of Scripture (Ps. 19:7-11). Any appeal to an unbeliever could be our last. What a shame it would be to end a spiritual discussion without planting the seed of our Lord who was crucified (1 Cor. 2:2-3).

Might they laugh? Yes. Could they become angry, hostile or reject the message? Certainly; but the seed will have been planted. A casual exchange with a skeptic does not necessitate a lengthy exegesis of complicated sections of Scripture. You can finish a discussion by simply expressing faith in the God-man Jesus, His sacrificial death, resurrection, and quoting a verse from the Lord; thus, planting a seed of faith (1 Cor. 3:5-8).

"But, what about Paul on Mars Hill?"

Some Christians argue that because Paul didn't quote Scripture in Athens (Acts 17); therefore, we shouldn't use Scripture to teach unbelievers. It's true that Paul began teaching the Athenians at their level of understanding. He discussed their altars (vs. 23), and the universal desire to seek God (vvs. 26, 27), as affirmed by their own poets (vs. 28); yet, how did Paul conclude his message?

"Truly, these times of ignorance God overlooked, but now commands all men everywhere to repent, because He has appointed a day on which He will judge the world in righteousness by the Man whom He has ordained. He has given assurance of this to all by raising Him from the dead" (Acts 17:30-31).

Paul did not shrink from declaring the gospel, even to those he knew would laugh: "And when they heard of the resurrection of the dead, some mocked while others said, 'We will hear you again on this matter.' So Paul departed from among them" (Acts. 17:32,33).

Paul did two important things: (1) He planted seeds from God's word, and (2) he immediately influenced others to faith in the Lord:"However, some men joined him and believed, among them Dionysius the Areopagite, a woman named Damaris, and others with them" (Acts 17:34).

Nevertheless, even *if* none of the Athenians ever obeyed the gospel, Paul would still have fulfilled his mission and glorified Jehovah in the process.

"But, what will unbelievers think if we tell them about the cross before they believe in the existence of God?"

Most will probably sneer (as countless messengers of the Lord have been rejected for millennia—cf. Heb. 11:35-38), but at least the seed of God will be planted for future growth. And for those who never believe?...

" ...The message of the cross is foolishness to those who are perishing, but to us who are being saved it is the power of God. For it is written: 'I will destroy the wisdom of the wise, and bring to nothing the understanding of the prudent.' Where is the wise?. .Where is the disputer of this age? Has not God made foolish the wisdom of this world?... it pleased God through the foolishness of the message preached to save those who believe... Because the foolishness of God is wiser than men, and the weakness of God is stronger than men... not many wise according to the flesh, not many mighty, not many noble, are called. But God has chosen the foolish things of the world to put to shame the wise... to bring to nothing the things that are, that no flesh should glory in His presence. But of Him you are in Christ Jesus' " (I Cor. 1:18-30).

Conclusion

There are seven billion people on earth today (10 billion by 2050), providing no shortage of fields white for harvest, should someone reject the message. Let's provide a healthy apologetic, leveraging the best scientific, archaeological, and historical data available, and plant the word of God in the process (Rom. 10:17).

Acknowledgments

Thanks to Lee Bailey, Trevor Bowen, Dan Bunting, Jim Deason, David Halbrook, Scott Long, Phil Martin, Bob Myhan, David Posey, Joe Price, Chris Reeves, Joe Works, and David Watts Jr. for providing helpful feedback and corrections.

Appendix D

Evolutionists who Admit the Geological Column Was Formed by Catastrophic Event(s) Rather Than Uniformitarian (Slow and Gradual) Processes

Dr. Warren Allmon, Director of the Paleontological Research Institution in Ithaca, NY, and Adjunct Associate Professor of Earth and Atmospheric Sciences at Cornell University—"Lyell also sold geology some snake oil. He convinced geologists that because physical laws are constant in time and space and current processes should be consulted before resorting to unseen processes, it necessarily follows that all past processes acted at essentially their current rates (that is, those observed in historical time). This extreme gradualism has led to numerous unfortunate consequences, including the rejection of sudden or catastrophic events in the face of positive evidence for them, for no reason other than that they were not gradual" (Allmon, 1993).

Dr. James H. Shea—"If the creationists could mount a successful attack on the validity of uniformitarianism, they would succeed in their effort to discredit modern geology" (Shea, 1983).

Dr. Kenneth J. Hsu (Chairman of the Experimental Geology Department at the Swiss Federal Institute of Technology, Zurich, where he was Emeritus Professor of Geology) and Dr. Judith A. McKenzie (Emeritus Professor of Geology, Swiss Federal Institute of Technology Zurich)—"Catastrophism is enjoying a renaissance in geology. For the last 180 years, geologists have applied consistently a uniformitarian approach to their studies that has stressed slow gradual changes as defined by Lamarck, Lyell, and Darwin. Now, many of us are accepting that unusual catastrophic events have occurred repeatedly during the course of Earth's history" (Hsu and McKenzie, 1986).

Dr. David M. Raup (University of Chicago Paleontologist, Curator and Dean of Science at the Field Museum of Natural History in Chicago)—"A great deal has changed, however, and contemporary geologists and paleontologists now generally accept catastrophe as a 'way of life' although they may avoid the word catastrophe" (Raup, 1983).

Dr. Robert H. Dott, Jr. (University of Wisconsin, Stanley A. Tyler Distinguished Professor of Sedimentary Geology)—"I hope I have convinced you that the sedimentary record is largely a record of episodic events rather than being uniformly continuous. My message is that episodicity is the rule, not the exception. "What do I mean by 'episodic sedimentation?' Episodic was chosen carefully over other possible terms. 'Catastrophic' has become popular recently because of its dramatic effect, but it should be purged from our vocabulary because it feeds the neo-catastrophist-creation cause" (Dott, 1982).

Dr. Digby J. McLaren (Canadian geologist and paleontologist, Director of the Geological Survey of Canada)—"A new uniformitarianism has moved to embrace at least a modified form of catastrophism" (McLaren, 1987).

Dr. U.B. Marvin (Senior Geologist, Emerita of Geology and Historian of Science at the Harvard-Smithsonian Center for Astrophysics)—"But to regard the cataclysmic geologic effects of bolide impacts as uniformitarian is an exercise in 'new speak,' whereby we would impose a 1980's usage on an 1830's term, which since the time it was coined, has donated the exact opposite of cataclysmic."

"Rather than to invert the definition of the venerable word, it is time to recognize that bolide impact is a geologic process of major importance, which by its very nature demolishes uniformitarianism itself as the basic principle of geology" (Marvin, 1990).

Dr. Stephen J. Gould (renowned Harvard paleontologist)—"Uniformitarianism is a dual concept. Substantive uniformitarianism (a testable theory of geologic change postulating uniform, pity of rates of material conditions) is false and stifling to hypothesis formation... Substantive uniformitarianism as a descriptive theory has not withstood the test of new data and can no longer be maintained in any strict manner... As a special term, methodological uniformitarianism was useful only when science was debating the status of the supernatural in its realm; for if God intervenes, then laws are not invariant and induction becomes invalid... The term today is an anachronism: [outdated term] for we need no longer take special pains to affirm the scientific nature of our discipline" (Gould, 1965).

"I wish to argue the following: 1) Gradualism has operated for the past one hundred and fifty years as a serious constraining bias in the history of geology. 2) Gradualism was never 'proved from the rocks' by Lyell and Darwin, but was rather imposed as a bias upon nature" (Gould, 1984, p. 16).

Dr. Peter E. Gretener (professor, Department of Geology, University of Calgary)—"Despite many rescue attempts, the term 'uniformitarianism' remains an unfortunate choice. Overwhelming evidence demonstrates that the course of the earth's history is anything but uniform. The term 'uniformitarianism" should be abolished because it is misleading" (Gretener, 1984, p. 87).

Dr. R.H. Dott, Stanley A. Tyler Distinguished Professor of Sedimentary Geology, University of Wisconsin—"Finally, it even means that catastrophism, in the sense of not straining the intensities of processes, was a better premise than Lyell's uniformitarianism" (Dott, 1998).

Dr. Victor R. Baker, Regents Professor, Department of Hydrology and Water Resources, Joint Professor, Geosciences; Joint Professor, Planetary Sciences, University of Arizona—"It has been the thesis of this essay that the 'new catastrophism' is rooted in a very old idea, one held by many of the old catastrophists: geology is about what Earth has to say to us" (Baker, 1998).

Dr. Derek V. Agar (evolutionist, professor and head of the Department of Geology and Oceanography, University College of Swansea)—"Palaeontologists cannot live by uniformitarianism alone. This may be termed the Phenomenon of the Fallibility of the Fossil Record" (1993a, p26).

"The periodic catastrophic event may have more effect than vast periods of gradual evolution. This may be called the Phenomenon of Quantum Sedimentation" (1993a, p. 57).

"It seems to me that the stratigraphical record is full of examples of processes that are far from 'normal' in the usual sense of the word. In particular we must conclude that sedimentation in the past has often been very rapid indeed and very spasmodic. This may be called the 'Phenomenon of the Catastrophic Nature of the Stratigraphic Record'" (1993a, p. 70).

"The hurricane, the flood, or the tsunami may do more in an hour or a day than the ordinary processes of nature have achieved in a thousand years. Given all the millennia we have to play with in the stratigraphical record, we can expect our periodic catastrophes to do all the work we want of them" (1993a, pp. 68-69).

"My intention is to show that catastrophism (in my sense) or at least episodicity, is apparent in everything... For a century and a half the geological world has been dominated, one might even say brain-wahsed, by the gradualistic uniformitarianism of Charles Lyell. Any suggestion of 'catastrophic' events has been rejected as old-fashioned, unscientific and even laughable" (Ager, 1993b, p. xi).

"I must emphasize that I am concerned with the whole history of the earth and its life and in particular with the dangerous doctrine of uniformitarianism" (Ager, 1993b, p. xvi).

"I prefer to retain the word 'catastrophism,' if only because it is more evocative and represents more clearly the antithesis of Lyell's boring doctrine... The present generation of earth scientists has become aware that the history of the biosphere is not only one of gradual and stately changes but that it is accentuated by events of various kinds and degrees, most of which are so rare that they refute a uniformitarian approach" (Ager, 1993b, p. xvii).

"One must constantly ask oneself, 'Is the present a long enough key to unlock the secrets of the past?' I shall discuss this further in chapter twelve... I think that the 'catastrophist' Georges Cuviere was a better geologist than the 'uniformitarian' Charles Lyell and my first chapter is devoted to a defence of that great Frenchman" (Ager, 1993b, p. xviii).

"Coal seams, like so many other things in earth history, were formed in very brief moments, geologically speaking" (Ager, 1993b, p. 47).

"I find nothing in earth history more convincing of the catastrophic/episodic nature of the record than that of ancient terrestrial deposits" (Ager, 1993b, p. 64).

"So there was never anything gradual or continuous about igneous activity, either volcanic or plutonic and here surely, I am entitled to use the term 'catastrophism'" (Ager, 1993b, p. 163).

"The basic principle of uniformitarianism is, of course, that we can use the processes going on at the present time to interpret the events of the geological past... It may be that it is a very odd and atypical present that we have to use to try to understand the past... Looked at the other way round, we must ask ourselves if our present is really all that typical and we must always accept the basic constraint that it may be a very odd period in which we now live. Everyone knows, or at least accepts, that things were very different in the early days of earth history. There surely our strict uniformitarianism approach does not apply" (Ager, 1993b, pp. 165-166).

"One just cannot have a gradual earthquake, a gradual hurricane, a gradual storm surge, a gradual magnetic reversal, a gradual volcanic explosion or—for that matter—a gradual impact" (Ager, 1993b, p. 196).

"In the late Carboniferous Coal Measures of Lancashire, a fossil tree has been found, thirty-eight feet high and still standing in its living position. Sedimentation must therefore have been fast enough to bury the tree and solidify before the tree had time to rot" (Ager, 1993a, p. 65).

Appendix E

ICR's Fossil Analyses with Verified Original Soft Tissues and Alleged Evolutionarily Ages

Articles Published in Peer-Reviewed Journals (supplemental material supporting papers at dinosaursofttissue.com)

From the Institute for Creation Research   icr.org

#	Date	Description	Age	Publication
1	6/14/1992	Osteocalcin in a seismosaur bone	150MY	Muyzer, et al., Geology, 20:871-874
2	9/25/1992	DNA in amber	30MY	Morell, et al., Science, 257:1860
3	6/16/1994	Unaltered amino acids in amber insects	130MY	Bada, et al. Geochemica et Cosmochemica Acta, 58(14):3131-3135
4	6/16/1994	Dinosaur DNA from hadrosaur bone	65MY	Woodward, et al., Science, 266(5188):1229-1232
5	5/19/1995	Live bacteria spores from amber	25-40MY	Cano and Borucki, Science, 268(5213):1060-1064
6	6/10/1997	Hemoglobin fragments in T. rex bone	67MY	Schweitzer, et al., PNAS, 94(12):6291-6296
7	6/2/1999	Live bacteria from halite [salt] deposit	250MY	Vreeland, et al., Amer. Soc. For Microbiology, 99th General Mtg.
8	6/21/1999	Live bacteria from separate rock salts	250MY	Stan-Lotter, et al. Microbiology, 145(12):3565-3574
9	6/21/1999	Icthyosaur skin	19MY	Linghan-Soliar, et al., Proc. Royal Soc. B, 266(1436):2367-2373
10	6/21/1999	Keratin in Madagascar Cretaceous bird	65MY	Schweitzer, et al., Jour. Vertebrate Paleontology, 19(4):712-722
11	9/1/2001	T. rex collagen SEM scans	65MY	Armitage, Creation Research Society Quarterly, 38(2):61-66
12	6/26/2004	Live (non-spore) bacteria in amber	120MY	Greenblatt, et al., Microbial Ecology, 48(1):120-127
13	3/24/2005	T. rex soft tissue	68MY	Schweitzer, et al., Science, 307(5717):1952-1955
14	6/30/2007	T. rex collagen	68MY	Schweitzer, et al., Science, 316(5822):277-280
15	4/7/2008	Psittacosaurus skin	125MY	Lingham-Soliar, T. et al, Proc. Royal Soc. B, 275:775-780
16	7/8/2008	Feather melanocytes	100MY	Vinther, J. et al, Biology Letters, 6(1):128-131
17	4/30/2009	Hadrosaur blood vessels	80MY	Schweitzer, M. et al, Science, 324(5927):626-631
18	8/26/2009	Purple Messel feather nanostructure	40MY	Vinther, J.et al, Biology Letters, 6(1):128-131
19	5/19/2009	Primate "Ida" soft body outline	40MY	Franzen, J.L. et al, PLoS ONE, 4(5):e5723
20	7/1/2009	Hadrosaur skin cell structures	66MY	Manning, P. et al, Proc. Royal Soc. B, 276:3429-3437
21	8/18/2009	Squid ink	150MY	Whilby, P.R. et al, Geology Today, 24(3):95-98
22	10/2/2009	Fungal chitin ubiquitous in Permo-triassic	250MY	Jin, Y. G. et al, Science, 289(5478):432-436
23	10/14/2009	Soft frog, intact, with bloody bone marrow	10MY	McNamara, et al., Geology, 34(8):641-644
24	11/5/2009	Salamander muscle, whole	18MY	McNamara, M. et al, Proc. Royal Soc. B, 277(1680):423-427
25	2/25/2010	Sinosauropteryx melanosomes	125MY	Zhang, F. et al, Nature, 463:1075-1078
26	3/10/2010	Psittacosaurus skin color	125MY	Linghan-Soliar, T.G. and Plodowski, Naturwissenschaften, 97:479-486
27	5/14/2010	Mammal hair in amber	100MY	Vullo, R., Naturwissenschaften, 97(7):683-687
28	5/18/2010	Archaeopteryx original biological material	150MY	Bergmann, U., PNAS, 107(20):9060-9065
29	8/9/2010	Mosasaur blood, retina	65-68MY	Lindgren, J., PLoS ONE, 5(8); e11998
30	11/12/2010	Penguin feathers [fossilized organelles]	36MY	Clarke, J.A. 35 al, Science, 330:954-957
31	11/18/2010	Shrimp shell and muscle	360MY	Feldman, R.M. and C.E. Schweitzer, J. Crustacean Biology, 30(4):629-635
32	2/7/2011	Chitin and chitin-associated protein	417MY	Cody, G.D. et al, Geology, 39(3):255-258
33	4/1/2011	C-14 date of mosasaur (24,600 Yrs)	70MY	Lindgren, J. et al, PLoS ONE, 6(4):e19445
36	3/23/2011	Lizard tail skin, Green River	40MY	Edwards, N. P. et al, Proc Royal Soc B, online
37	6/8/2011	Type I Collagen, T.rex and hadrosaur	68MY	San Antonio, J.D. et al, PLoS ONE, 6(6):e20381
38	6/30/2011	Bird feather pigment	120MY	Wogelius, R.A. et al, Science, online
		Preliminary Reports Published Elsewhere		
39	8/10/2009	Live yeast in amber	45MY	Wired Science
40	4/10/2010	Australopithecus sediba brains	1.9MY	Discovery News
41	9/27/2010	Lobster shell	"millions"	Keighley News
42	10/22/2010	Mosasaur cartilage	80MY	Buchholz, C.C., Rapid City Journal
		See Also: Publications added by Real Science Radio		
	1993-2013	Dino and dino-layer biological material	65-150MY	Science, J Vert Pal, J Ap Genet, Proc R Soc, PLoS One, PNAS, Nature, etc.
	1964	??	??	Kabat, E. A. et al, Experimental Immunochemistry 2nd ed., pp. 85-90

1966	Collagen & vessels in dinosaur bone	200MY	Pawlicki R., et al, Cells, collagen fibrils and vessels in dinosaur bone, Nature
1968	Collagen from Megalosaurus egg shell	166MY	Voss-Foucart, Paleoproteins, Comparative Biochemistry and Physiology
1968	Intact proteins from half of many bones	150MY	Miller & Wyckoff, Proteins in Dinosaur Bones, Proc. Nat'l Acad. of Sciences
1972	The biochemistry of animal fossils	??	Wyckoff, R. W. G., Scientechnica Ltd, Bristol
1972	Biological material in saltwater clam shell	??	Crenshaw, M. A., Biomineralization Research Reports, 6:6–11
1973	??	??	Westbroek, P., et al, Calcified Tissue Research, 12:227–238
1974	Proteins and polysaccharides	70MY	DeJong, E. W., et al, Nature 252:63–64
1985	Metabolic pathways of dinosaur bones	80MY	Pawlicki R. Metabolic pathways of fossils. Part V. Histochem Cytochem
1987	Seven hadrosaurs' unfossilized bones	80MY	Davies K. Duck-billed dinosaurs from the north slope of Alaska, J of Paleont.
			Ostrom, P.H. et al, Geology, Assessment of tropic structure of
1993	Organic material in dinosaur specimens	65MY+	Creataceous...
1993	Chloroplast tree gene partly sequenced	35-40MY	Poinar, H.N. et al, Nature correspondence, 363(6431):677
2008	Osteocyte (bone cells)	5MY	Bell LS, et al, Mineralized osteocyte: a living fossil. Am J Phys 137:449-56
2012	Turtle osteocytes	145MY+	Cadena E. Schweitzer MH. Bone 51:614-620
2013	Signature of blood in mosquito gut	46MY	Greenwalt, D., et al, Hemoglobin-derived porphyrins. PNAS Oct. 14
2013	Biological material in crinoids	350MY	O'Malley, et al, Taxon-specific organic molecules, Geology 41(3):347-350
in preparation	Hadrosaur blood vessels	80MY	Cleland, TP, Zheng W, Zamdborg L, Schweitzer MH. Molecular charact.
2014	Precambrian Metaoans	530MY	Moczydłowska, M., et al, Ediacaran Metozoa, J of Paleontology 88(2):224+

Please email suggested additions, corrections, fill-in-the-blanks, and clarifications to Bob@RealScienceRadio.com.

Correction: Dr. Whitmore (Cedarville creationist associate professor of geology) noticed that no biological material was identified for what was
RSR Update: The penguin feathers had fossilized 'organelles' (for defintion, listen to rsr.org/organelles). We're going to review that paper to consider

Bill Nye said that creationism makes no predictions.   But see another RSR prediction confirmed! This time, by allegedly 530 mya soft tissue!

Appendix F

Table 1 from Baumgardner (2005), Displaying Peer-Reviewed Publications Detecting Carbon-14 in Specimens Supposedly Millions of Years Old

Table 1. Accelerator mass spectrometer measurements for Percent Carbon-14 (C-14) on samples conventionally deemed older than 100,000 years.

(NOTE: C-14 is undetectable after 40,000 years (Bowman, 1990) due to a half-life of 5,730 years. Thus, these samples that are supposed to be millions of years old are shown to be less than 100,000 years old. Baumgardner (2005) explains,"Values ranged from 7.58±1.11 percent of the modern atmospheric $^{14}C/C$ ratio (commonly referred to as percent modern carbon, or pMC) for a lower Jurassic sample to 0.38±0.04 pMC for a middle Tertiary sample. This range in $^{14}C/C$ ratio implies radiocarbon ages of between 20,700±1200 and 44,700±950 years).

Table 1 modified from Baumgardner (2005).

	$^{14}C/C$ (pMC) (±1 Standard Deviation)	Material	Reference
1	0.71±?*	Marble	*Aerts-Bijma et al.* [1997]
2	0.65±0.04	Shell	*Beukens* [1990]
3	0.61±0.12	Foraminifera	*Arnold et al.* [1987]
4	0.60±0.04	Commercial graphite	*Schmidt et al.* [1987]
5	0.58±0.09	Foraminifera *(Pyrgo murrhina)*	*Nadeau et al.* [2001]
6	0.54±0.04	Calcite	*Beukens* [1990]
7	0.52±0.20	Shell *(Spisula subtruncata)*	*Nadeau et al.* [2001]
8	0.52±0.04	Whale bone	*Jull et al.* [1986]
9	0.51±0.08	Marble	*Gulliksen and Thomsen* [1992]
10	0.5±0.l	Wood, 60ka	*Gillespie and Hedges* [1984]
11	0.46±0.03	Wood	*Beukens* [1990]
12	0.46±0.03	Wood	*Vogel et al.* [1987]
13	0.44±0.13	Anthracite	*Vogel et al.* [1987]
14	0.42±0.03	Anthracite	*Grootes et al.* [1986]
15	0.401±0.084	Foraminifera (untreated)	*Schleicher et al.* [1998]
16	0.40±0.07	Shell *(Turitella communis)*	*Nadeau et al.* [200 l]
17	0.383±0045	Wood (charred)	*Snelling* [1997]
18	0.358±0033	Anthracite	*Beukens et al.* [1992]

19	035±003	Shell *(Varicorbula gibba)*	*Nadeau et al.* [2001]
20	0.342±0.037	Wood	*Beukens et al.* [1992]
21	0.34±0.11	Recycled graphite	*Arnold et al.* [1987]
22	0.32.±0.06	Foraminifera	*Gulliksen and Thomsen* [1992]
23	0.3±?	Coke	*Terrasi et al.* [1990]
24	0.3±?	Coal	*Schleicher et al.* [1998]
25	0.26±0.02	Marble	*Schmidt et al.* [1987]
26	0.2334±0.061	Carbon powder	*McNichol et al.* [1995]
27	0.23±0.04	Foraminifera (mixed species avg)	*Nadeau et al.* [2001]
28	0.211±0.018	Fossil wood	*Beukens* [1990]
29	0.21±0.02	Marble	*Schmidt et al.* [1987]
30	0.21±0.06	CO_2	*Grootes et al.* [1986]
31	020-0.35* (range)	Anthracite	*Aerts-Bijma et al.* [1997]
32	0.20±004	Shell *(Ostrea edulis)*	*Nadeau et al.* [2001]
33	0.20±004	Shell *(Pecten opercularis)*	*Nadeau et al.* [2001]
34	0.2±0.1*	Calcite	*Donahue et al.* [1997]
35	0.198±0.060	Carbon powder	*McNichol et al.* [1995]
36	0.18±0.05 (range?)	Marble	*Van der Borg et al.* [1997]
37	0.18±0 03	Whale bone	*Gulliksen and Thomsen* [1992]
38	0.18±0 03	Calcite	*Gulliksen and Thormen* [1992]
39	0.18±0.01**	Anthracite	*Nelson et al.* [1986)
40	0.18±?	Recycled graphite	*Van der Borg et al.* [1997]
41	0.17±0.03	Natural gas	*Gulliksen and Thomsen* [1992]
42	0.166±0.008	Foraminifera (treated)	*Schleicher et al.* [1998]
43	0.162±?	Wood	*Kirner et al.* [1997]
44	0.16±0 03	Wood	*Gulliksen and Thomsen* [1992]
45	0.154±?**	Anthracite coal	*Schmidt et al.* [1987]
46	0.152±0.025	Wood	*Beukens* [1990]
47	0.142±0.023	Anthracite	*Vogel et al.* [1987]
48	0.142±0.028	CaC_2, from coal	*Gurfinkel* [1987]
49	0.14±0.02	Marble	*Schleicher et al.* [1998]
50	0.13±0.03	Shell *(Mytilus edulis)*	*Nadeau et al.* [2001]
51	0.130±0.009	Graphite	*Gurfinkel* [1987]
52	0.128±0.056	Graphite	*Vogel et al.* [1987]
53	0.125±0.060	Calcite	*Vogel et al.* [1987]

54	0.12±0.03	Foraminifera (N. pachyderma)	Nadeau et al. [2001]
55	0.112±0.057	Bituminous coal	Kitagawa et al. [1993]
56	0.1±0.01	Graphite (NBS)	Donahue et al. [1990]
57	0.1±0.05	Petroleum, cracked	Gillespie & Hedges [1984]
58	0.098±0.009*	Marble	Schleicher et al. [1998]
59	0.092±0.006	Wood	Kirner et al. [1995]
60	0.09-0.18* (range)	Graphite powder	Aerts-Bijma et al. [1997]
61	0.09-0.13* (range)	Fossil CO_2, gas	Aerts-Bijma et al. [1997]
62	0.089±0.017	Graphite	Arnold et al. [1987]
63	0.081±0.019	Anthracite	Beukens [1992]
64	0.08±?	Natural Graphite	Donahue et al. [1990]
65	0.080±0.028	Carrarra marble	Nadeau et al. [2001]
66	0.077±0.005	Natural Gas	Beukens [1992]
67	0.076±0009	Marble	Beukens [1992]
68	0.074±0.014	Graphite powder	Kirner et al. [1995]
69	0.07±?	Graphite	Kretschmer et al. [1998]
70	0.068±0.028	Calcite (Icelandic double spar)	Nadeau et al. [2001]
71	0.068±0.009	Graphite (fresh surface)	Schmidt et al. [1987]
72	0.06- 0.11 (range)	Graphite (200 Ma)	Nakai et al. [1984]
73	0.056±?	Wood (selected data)	Kirner et al. [1997]
74	0.05±0.01	Carbon	Wild et al. [1998]
75	0.05±?	Carbon-12 (mass spectrometer)	Schmidt, et al. [1987]
76	0.045-0.012 (-0.06)	Graphite	Grootes et al. [1986]
77	0.04±?*	Graphite rod	Aerts-Bijma et al. [1997]
78	0.04±0.01	Graphite (Finland)	Bonani et al. [1986)
79	0.04±0.02	Graphite	Van der Borg et al. [1997]
80	0.04±0.02	Graphite (Ceylon)	Bird et al. [1999]
81	0.036±0.005	Graphite (air)	Schmidt el al. [1987]
82	0.033±0.013	Graphite	Kirner et al. [1995]
83	0.03±0.015	Carbon powder	Schleicher et al. [1998]
84	0.030±0.007	Graphite (air redone)	Schmidt e1 al. [1987]
85	0.029±0.006	Graphite (argon redone)	Schmidt et al. [1987]
86	0.029±0.010	Graphite (fresh surface)	Schmidt et al. [1987]
87	0.02±?	Carbon powder	Pearson et al. [1998]
88	0.019±0.009	Graphite	Nadeau et al. [2001]
89	0.019±0.004	Graphite (argon)	Schmidt el al. [1987]
90	0.014±0.010	CaC_2, (technical grade)	Beukens [1993]

* Estimated from graph ** Lowest value of multiple dates

Appendix G

Closing the Door on the Non-Human Neanderthal Myth

Introduction

It was once assumed that Neanderthal man was a missing link in human evolution. Now that evolutionists themselves have thoroughly debunked that idea, public schools continue to teach our most vulnerable young minds that Neanderthal man was an unintelligent, non-human, cave-man, hominid race—related to man by common ancestors. One children's book states, "About two million years ago, ancient humans first traveled out of Africa. Soon they had journeyed east to Java and China, and by over a million years ago their descendants had also gotten into Europe where they eventually gave rise to *Homo neanderthalis*" (i.e., "Neanderthal man", JBG) (1). What is the impact of this statement? Neo-Darwinian evolution teaches that humans and Neanderthals branched off from a common ancestor hundreds of thousands of years ago and evolved into separate species. **Translation:** The account of Adam and Eve is a myth. Man supposedly evolved in the following scheme:

Chimpanzee → *Ardipithecus* → *Australopithecus afarensis* → *Kenyanthropus platyops* → *Homo habilis* → *Homo erectus* → *Homo antecessor* → *Homo heidelbergensis* → modern *Homo sapiens sapiens* (humans).

Such a scheme effectively mythologizes the Genesis account of Adam and Eve, leading some children to conclude that the textbooks are fact and the Bible is a myth.

Disney-sponsored Bill Nye the Science Guy teaches kids, "Now this process, this thing that happens is called evolution, and it's been going on for billions of years" (2). "Man and apes came from the same stem of the evolutionary tree, but man branched off and descended, while the apes degenerated... The animals that

we descended from that lived millions of years ago used to be able to swing through trees on their tails" (3). These types of unfounded assertions are considered gospel to so many secularists and their pupils today; in addition to some believers in Christ, who have compromised the Biblical account of creation (Genesis 1 & 2).

Human or ape: there's no "Mr. In-between"

The taxonomic family called Hominids includes both humans and the great apes. We are told that this family also includes intermediate ape-man creatures that humans evolved from—our "ancestors." In addition, there were non-human creatures such as the unintelligent, boorish, cave-man Neanderthals who, although they were not in direct descent to modern man, shared many human-like qualities. On the other hand, creationists contend that there are no ape-men, or unevolved cave-men, and that all hominids fall into only one of two distinct categories. Namely, all hominids are either (I) true apes or (II) true humans. Where does that leave Neanderthal? I contend that Neanderthal was 100% human.

Reasons to Believe that Neanderthals were True Humans

1. Coexistence with humans and the same species as humans. Neanderthals did not live prior to humans, but coexisted with humans. No evolutionist disputes this point. Despite the children's books, some secular scientists now, quite surprisingly, even classify Neanderthals as a subspecies of *Homo sapiens* (viz., *Homo sapiens neanderthalensis*), which is the same species as humans. In other words, some evolutionists teach that Neanderthals and humans were the very same species (4-7).

2. What about DNA? A groundbreaking fifty-five author article in *Science* compared the genomes of three Neanderthals to the genomes of five present-day individuals. They concluded (based on their DNA analyses) that humans and Neanderthals were highly genetically similar, and even stated, "Neandertals are expected to be more closely related to some present-day humans than they are to each other... The majority of the Neandertal divergences overlap with those of the humans, reflecting the fact that Neandertals fall inside the variation of present-day humans." (8) In other words, Neanderthals were genetically within the range of humans living today!

3. Interbreeding with humans. Secular scientists have also slowly come around to admitting, with the creationists, that Neanderthals and humans interbred with one another. So says the prestigious scientific journal: *Nature* (9, 10). Doesn't it sound like Neanderthals were human to you?

4. Structurally-similar to humans. It has been known since the 1960's that the anatomical measurements of Neanderthal man are well within the range of measurements derived from living humans today.

5. Burying their dead. Neanderthals buried their dead with ritualistic honors including flowers. One Neanderthal skeleton is buried along with an elephant tusk, and a Neanderthal child is buried with mountain goat horns ritualistically surrounding the body. Neanderthals maintained family burial grounds in caves or rock shelters in at least 258 sites discovered. At four sites, humans and Neanderthals were found buried *together* (11). That's right—together. (NOTE: Humans are the only creatures known to bury their dead, and would likely not do so along with some archaic, unintelligent, non-human "cave men," with whom they could not communicate).

These burial rituals are similar to those described in the Bible in Genesis. People buried in caves included Sarah (Gen. 23:17–20), Abraham (Gen. 25:7–11), and Jacob (Gen. 49:29–32). This practice took place in New Testament times as well, as both Lazarus and Jesus were buried in caves (John 11:38; Matt. 27:60).

6. Miscellaneous reasons to believe that Neanderthals were human. Other evidence that indicates the Neanderthals were fully human include their grinding of grain for food, setting broken bones, performing surgery, making ornate jewelry, making musical instruments such as a flute constructed from animal bone, making intricate tools, such as a lissoir bone tool used for working leather, butchering their own meat with knives, hunting and butchering elephants and wooly mammoths, using 6-7.5 foot throwing spears, making a human facemask out of flint, controlled use of fire for cooking and tool working, constructing structured, walled living areas, and separating litter debris from clear ground. Some Neanderthal tools even had handles inserted in them, sealed with an adhesive (bitumen), which was cured at high temperatures (11).

Conclusion

So, who were the Neanderthal? Based on their burial grounds not being immersed in flood sediment, they were probably a genetic variant of humans that spread out and descended from the eight people who survived the Noahic flood. They may have been a tribe that separated off after the languages were confused at Babel. Because they lived closer to Adam, in the post-flood era, their life spans may have been significantly longer than ours today (NOTE: Gen. 9:29 —Noah lived to be 950 years old). This could have contributed to the heavy-boned features of the race, such as their thick elongated skulls and protruding clavicle bones and ridge brows. Let's recap: Neanderthals acted like humans, interacted with humans, interbred with humans, and were buried with humans in death. So, why not simply conclude that Neanderthals are humans? Unless one has an ulterior motive (viz., Neo-Darwinism).

Originally published as:

Gurtler, Joshua (2015), "Closing the Door on the Non-Human Neanderthal Myth," *Biblical Insights Magazine,* 15[4]:14-15 , April.

References

1. Tatersall, I., and R. DeSalle. 2013. The Great Human Journey: Around the World in 22 Million Days. Bunker Hill Pub: Piermont, NH. http://www.amazon.com/The-Great-Human-Journey-Million/dp/1593731485

2. Bill Nye the Science Guy- Evolution (1/2). Youtube. Accessed on Jan. 24, 2015 at: https://www.youtube.com/watch?v=svHQ4BQY__o

3. Bill Nye the Science Guy- Evolution (2/2). Youtube. Accessed on Jan. 24, 2015 at: https://www.youtube.com/watch?v=QECq6M3nPew

4. The Neandertal lower right deciduous second molar from Trou de l'Abîme at Couvin, Belgium. Toussaint M, Olejniczak AJ, El Zaatari S, Cattelain P, Flas D, Letourneux C, Pirson S. J Hum Evol. 2010 Jan;58(1):56-67. http://www.ncbi.nlm.nih.gov/pubmed/19910020

5. Discovery of Homo sp. tooth associated with a mammalian cave fauna of Late Middle Pleistocene age, northern Thailand. Tougard C, Jaeger JJ. J Hum Evol. 1998 Jul;35(1):47-54 http://www.ncbi.nlm.nih.gov/pubmed/9680466

6. Human remains of **Homo sapiens neanderthalensis** from the pleistocene deposit of Santa [corrected] Croce Cave, Bisceglie (Apulia), Italy. Mallegni F, Piperno M, Segre A. Am J Phys Anthropol. 1987 Apr;72(4):421-9. Erratum in: Am J Phys Anthropol 1988 Jan;75(1):143. http://www.ncbi.nlm.nih.gov/pubmed/3111268

7. Paleoclimatic setting for **Homo sapiens neanderthalensis**. Boaz NT, Ninkovich D, Rossignol-Strick M. Naturwissenschaften. 1982 Jan;69(1):29-33. http://www.ncbi.nlm.nih.gov/pubmed/6799843

8. A draft sequence of the Neandertal genome. Green RE *et al.* Science. 2010 May 7;328(5979):710-22. http://www.ncbi.nlm.nih.gov/pubmed/20448178

9. The genomic landscape of Neanderthal ancestry in present-day humans. Sankararaman S, Mallick S, Dannemann M, Prüfer K, Kelso J, Pääbo S, Patterson N, Reich D. Nature. 2014 Mar 20;507(7492):354-7. http://www.ncbi.nlm.nih.gov/pubmed/24476815

10. Sanchez-Quinto, F. *et al.* 2012. North African Populations Carry the Signature of Admixture with Neandertals. PLoS ONE. 7 (10): e47765. http://journals.plos.org/plosone/article?id=10.1371/journal.pone.0047765

11. The Neandertals: Our Worthy Ancestors, Part II. The Fossil and Archaeological Evidence. Marvin Lubenow. April 11, 2007. Accessed online on Jan. 24, 2015 at: https://answersingenesis.org/human-evolution/neanderthal/the-neandertals-our-worthy-ancestors-part-ii/

Appendix H

More quotes from mainstream evolutionists who say that Neo-Darwinism is dead.

Dr. Robert Wesson, Stanford University Hoover Institution Senior Research Fellow (as quoted by Fodor and Piattelli-Palmarini, 2010, p. 224)—"In a book that we think deserved greater attention and circulation than it received, the late political philosopher Robert Wesson wrote: 'Biologists, it seems, must do without a comprehensive theory of evolution, just as social scientists have to make do without a comprehensive theory of society' " (Wesson, 1991).

Lynn Margulis, evolutionist, American biologist and university professor in the Department of Geosciences at the University of Massachusetts Amherst; 2008 Darwin-Wallace Medalist; ex-wife of the late Carl Sagan—"The notion is that if we accumulate enough gene change, enough genetic mutations, we explain the passage from one species to another. This is depicted as two branches in a family tree that emerge from one common ancestor to the two descendants. An entire Anglophone academic tradition of purported devolutionary description was developed, quantified, computerized based on what I think is a conceptual topological error" (Margulis, 2010, p. 274).

"If, as I claim, heritable variation mostly does NOT come from gradual accumulation of random mutation, what does generate Darwin's variation upon which his natural selection can act? A fine scientific literature on this theme actually exists and grows every day but unfortunately it is scattered, poorly understood and neglected nearly entirely by the money-powerful, the publicity mongers of science and the media" (Margulis, 2010, p. 281).

"[Neo-Darwinism will one day be viewed as] "a minor twentieth century religious sect within the sprawling religious persuasion of Anglo-Saxon Biology"... [Neo-Darwinists] "wallow in their zoological, capitalistic, competitive, cost-benefit interpretation of Darwin... Neo-Darwinism, which insists on the slow accrual of mutations by gene-level natural selection, is a complete funk" (Margulis, 1991).

"The source of purposeful inherited novelty in evolution, the underlying reason the new species appear, is not random mutation" (Margulis, 2010, p. 279).

"New mutations don't create new species; they create offspring that are impaired... Darwinian claim to explain all of evolution is a popular half - truth whose lack of explicative power is compensated for only by the religious ferocity of its rhetoric" (Margulis, 2006).

"Mutations, in summary, tend to induce sickness, death, or deficiencies. No evidence in the vast literature of heredity changes shows unambiguous evidence that random mutation itself, even with geographical isolation of populations, leads to speciation" (Margulis and Sagan, 2002, p. 29).

Evolutionist Dr. Brian Goodwin, Professor of Biology, Open University, U.K.—"My main criticism of Darwinism is that it fails in its initial objective, which is to explain the origin of species. Now, let me explain exactly what I mean by that. I mean it fails to explain the emergence of organisms, the specific forms during evolution like algae and ferns and flowering plants, corals, starfish, crabs, fish, birds. ...Darwin turned biology into a historical science, and in Darwinism, species are simply accidents of history, they don't have any inherent nature. They are just 'the way things happened to work out' and there aren't any particular constraints that mean it couldn't have all worked out very differently" (Goodwin, 1996).

"Clearly something is missing from biology. It appears that Darwin's theory works for the small-scale aspects of evolution: it can explain the variations and the adaptations within species that produce fine-tuning of varieties to different habitats. The large-scale differences of form between types of organism that are the foundation of biological classification systems seem to require a principle other than natural selection operating on small variations, some process that gives rise to distinctly different forms of organism. This is the problem of emergent order in evolution, the origins of novel structures in organisms that has always been a primary interest in biology" (Goodwin, 1995a, p. xxi).

Dr. Egbert Leigh, Smithsonian Tropical Research Institute Staff Scientist Emeritus, Evolutionary Biologist, Ecologist—"The primary problem with the [modern evolutionary] synthesis is that its makers established natural selection as the director of adaptive evolution by eliminating competing explanations, not by providing evidence that natural selection among 'random' mutations could, or did, account for observed adaptation (Box 2). Mayr remarked, 'As these non-Darwinian explanations were refuted during the synthesis... natural selection automatically became the universal explanation of evolutionary change (together with chance factors).' Depriving the synthesis of plausible alternatives, which seemed such a triumph, in fact sowed the seeds of its faults" (Leigh, 1999).

Dr. Fred Hoyle - mathematician, physicist and Professor of Astronomy, Cambridge University—"What was in no way guaranteed by the evidence, however, was that evolutionary inferences correctly made in the small for species and their varieties could be extrapolated to broader taxonomic categories, to kingdoms, divisions, classes, and orders. Yet this is what the Darwinian theory did, and it was by going far outside its guaranteed range of validity that the theory ran into controversies and difficulties which have never been cleared up over more than a century" (Hoyle, 1999).

Peter Hitchens (brother of Christopher Hitchens)—"The BBC teased religious leaders by asking them if they believed in the literal truth of the great Bible stories. I would like to ask BBC chiefs and the rest of our secular establishment if they believe in the literal truth of evolution. Evolution is an unproven theory. If what its fundamentalist supporters believe is

true, fishes decided to grow lungs and legs and walk up the beach. The idea is so comically daft that only one thing explains its survival—that lonely, frightened people wanted to expel God from the Universe because they found the idea that He exists profoundly uncomfortable" (Hitchens, 2000).

Dr. Franklin M. Harold, Professor Emeritus of Biochemistry, Colorado State University—"We must concede that there are presently no detailed Darwinian accounts of the evolution of any biochemical or cellular system, only a variety of wishful speculations" *(Harold, 2001)*.

Evolutionist, atheist, and humanist, Rob Wipond—"But then, it merely exposes how much the belief in evolutionary theory is ultimately based upon a similar kind of blind faith. It shows there is no definitive, final proof for evolution, either. There are just a lot of suggestive facts that make some of us formulate an argument, every bit as tautological as the quote-the-Bible-to-prove-creationism-is-right arguments, which goes something like this: 'Evolution seems to have occurred; therefore, evolution has occurred' " (Wipond, 1998).

Dr. Eric H. Davidson, Norman Chandler Professor of Cell Biology, California Institute of Technology—"Neo-Darwinian evolution... erroneously assumes that change in protein-coding sequence is the basic cause of change in developmental program; and it erroneously assumes that evolutionary change in body plan morphology occurs by a continuous process. All of these assumptions are basically counterfactual" (2011).

"Contrary to classical evolution theory, the processes that drive the small changes observed as species diverge cannot be taken as models for the evolution of the body plans of animals. These are as apples and oranges, so to speak" (Davidson, 2006).

Dr. Keith Stewart Thomson, Emeritus Professor of Natural History at the University of Oxford, Director of the Oxford University Museum of Natural History—"The million-dollar question is: What mechanisms lie between the short-term, low-scale and wholly reversible results so far obtained, and the origin of a new species? What conditions and mechanisms are required to feedback from a given level of phenotypic plasticity to a new genetic or phenotypic constitution? Stay tuned" (Thomson, 1997).

Dr. Jerry Fodor, Ph.D. from Princeton University, Professor of Philosophy at Rutgers University—"Phenotypes aren't, in short, random collections of traits, and nonrandomness doesn't occur at random; the more nonrandomness there is, the less likely it is to have been brought about by chance. That's a tautology. So, if the nonrandomness of phenotypes isn't a reflection of the orderliness of God's mind, perhaps it is a reflection of the orderliness of the environments in which the phenotypes evolved. That's the theory of natural selection in a nutshell" (Fodor, 2007).

"The high tide of adaptationism [i.e. the modern evolutionary synthesis or Neo-Darwinism, JBG] floated a motley navy, but it may now be on the ebb. If it does turn out that natural selection isn't what drives evolution, a lot of loose speculations will be stranded high, dry and looking a little foolish. Induction over the history of science suggests that the best theories we have today will prove more or less untrue at the latest by tomorrow afternoon. In science, as elsewhere, 'hedge your bets' is generally good advice" (*Ibid.*).

"Anyhow, for what it's worth, I really would be surprised to find out that I was meant to be a hunter-gatherer since I don't feel the slightest nostalgia for that sort of life. I loathe the very idea of hunting, and I'm not all that keen on gathering either. Nor can I believe that living like a hunter-gatherer would make me happier or better. In fact, it sounds to me like absolute [expletive deleted, JBG]" (*Ibid.*).

"The years after Darwin witnessed a remarkable proliferation of other theories, each seeking to coopt natural selection for purposes of its own. Evolutionary psychology is currently the salient instance, but examples have been legion. They're to be found in more or less all of the behavioural sciences, to say nothing of epistemology, semantics, theology, the philosophy of history, ethics, sociology, political theory, eugenics and even aesthetics. What they have in common is that they attempt to explain why we are so-and-so by reference to what being so-and-so buys for us, or what it would have bought for our ancestors. [NOTE: Dr. Fodor then lists a number of evolutionary 'Just so' stories, which I will number sequentially.] (1)'We like telling stories because telling stories exercises the imagination and an imagination would have been a good thing for a hunter-gatherer to have.' (2)'We don't approve of eating grandmother because having her around to baby-sit was useful in the hunter-gatherer ecology.' (3)'We like music because singing together strengthened the bond between the hunters and the gatherers (and/or between the hunter-gatherer grownups and their hunter-gatherer offspring)'. (4)'We talk by making noises and not by waving our hands; that's because hunter-gatherers lived in the savannah and would have had trouble seeing one another in the tall grass.' (5)'We like to gossip because knowing who has been up to what is important when fitness depends on co-operation in small communities.' (6)'We don't all talk the same language because that would make us more likely to interbreed with foreigners (which would be bad because it would weaken the ties of hunter-gatherer communities).' (7)'We don't copulate with our siblings because that would decrease the likelihood of interbreeding with foreigners (which would be bad because, all else being equal, heterogeneity is good for the gene pool).' I'm not making this up, by the way. Versions of each of these theories can actually be found in the adaptationist literature. But, in point of logic, this sort of explanation has to stop somewhere. Not all of our traits can be explained instrumentally; there must be some that we have simply because that's the sort of creature we are. And perhaps it's unnecessary to remark that such explanations are inherently post hoc (Gould called them 'just so stories'); or that, except for the prestige they borrow from the theory of natural selection, there isn't much reason to believe that any of them is true" (*Ibid.*).

"it's not out of the question that a scientific revolution – no less than a major revision of evolutionary theory – is in the offing... The empirical issue. It wouldn't be unreasonable for a biologist of the Darwinist persuasion to argue like this: 'Bother conceptual issues and bother those who raise them. We can't do without biology and biology can't do without Darwinism. So Darwinism must be true.'... and it's entirely possible that adaptationism [i.e. the modern evolutionary synthesis or Neo-Darwinism JBG] is the wrong answer" (*Ibid.*)

"If, as I suggested, the notion of natural selection is conceptually flawed, such alternatives would be distinctly welcome... So what's the moral of all this? Most immediately, it's that the classical Darwinist account of evolution as primarily driven by natural selection is in trouble on both conceptual and empirical grounds" (*Ibid.*).

"Darwin was too much an environmentalist. He seems to have been seduced by an analogy to selective breeding, with natural selection operating in place of the breeder. But this analogy is patently flawed; selective breeding is performed only by creatures with minds, and natural selection doesn't have one of those... Darwinists do often argue this way" (*Ibid.*).

"A confusion between (I) the claim that evolution is a process in which creatures with adaptive traits are selected and (2) the claim that evolution is a process in which creatures are selected for their adaptive traits. We will argue that: Darwinism is committed to inferring (2) from (I); that this inference is invalid (in fact it's what philosophers call an 'intensional fallacy')" (Fodor and Piattelli-Palmarini, 2010, p. xv).

"Natural selection theory is often said to provide a mechanism for the evolution of phenotypes. That, however, is precisely what it doesn't do" (*Ibid.*, p. 148).

** " 'OK; so if Darwin got it wrong, what do you guys think is the mechanism of evolution?' Short answer: we don't know what the mechanism of evolution is. As far as we can make out, nobody knows exactly how phenotypes evolve" (*Ibid.*, p. 153).

"The point to keep your eye on is this: it is possible to imagine serious alternatives to the traditional Darwinian consensus that evolution is primarily a gradualistic process in which small phenotypic changes generated at random are then filtered by environmental constraints. This view is seriously defective" (*Ibid.*, p. 54).

"So far, these cases are not telling us that natural selection did not happen, but they tell us that the parallel between artificial selection and natural selection, so central to Darwin's theory is flawed" (*Ibid.*, p. 63).

** "The theory of natural selection is internally flawed; its' not just that the data are equivocal, it's that there's a crack in the foundations" (*Ibid.* p. 56).

"But, as we have just seen, there are some instances of optimal (or near-optimal) solutions to problems in biology; so, if natural selection cannot optimize, then something else must be involved. Very plausibly, the 'something else' includes: physics, chemistry, autocatalytic processes, dissipative structures and principles of self-organization, and surely other factors that the progress of science will in due time reveal" (*Ibid.*, p. 92)

"Adaptationism simply cannot do what an evolutionary theory is supposed to do: explain how phenotypic traits are distributed in populations of organisms. Equivalently: the theory of natural selection cannot predict/explain what traits the creatures in a population are selected-for" (*Ibid.*, p. 110).

** "So the claim that selection is the mechanism of evolution cannot be true... Advertising to the contrary nowithstanding, natural selection can't be a general mechanism that connects phenotypic variation with variation in fitness. So natural selection can't be the mechanism of evolution" (*Ibid.*, 114).

"In this respect Darwin was inadequately impressed by the fact that breeders have minds —they act out of their beliefs, desires, intentions and so on—whereas, of course, nothing of that sort of is true in the case of natural selection" (*Ibid.* p. xix).

** "We've been told by more than one of our colleagues that, even if Darwin was substantially wrong to claim that natural selection is the mechanism of evolution, nonetheless we shouldn't say so. Not, anyhow, in public. To do that is, however inadvertently, to align oneself with the Forces of Darkness, whose goal is to bring Science into disrepute. Well, we don't agree" (*Ibid.* p. xx).

** "In consequence, whereas Skinner's theory of conditioning is false, Darwin's theory of selection is empty" (*Ibid.* p. 16).

"These days biologists have good reasons to believe that selection among randomly generated minor variants of phenotypic traits falls radically short of explaining the appearance of new forms of life... We think of natural selection as tuning the piano, not as composing the melodies. That's our story, and we think it's the story that modern biology tells when it's properly construed" (*Ibid.* p. 21).

** "William Jeffery, an evolutionary developmental biologist at the University of Maryland, College Park, told Pennisi: 'You can collect lists of conserved genes, but once you get those lists, it's very hard to get at the mechanisms [of evolution]'. His conclusion is rather drastic: 'Macroevolution is really at a dead end' " (*Ibid.*, p. 30).

** "Erwin (2006) argued that known microevolutionary processes cannot explain the evolution of large differences in development that characterize entire classes of animals" (*Ibid.* 40).

** "The main thesis of this book is that NS [natural selection] is irredeemably flawed" (*Ibid.*, p. 1).

"This is a rigged game. The rule is: if a kind of creature fails to solve an evolutionary problem, it follows that that isn't an evolutionary problem for that kind of creature. Quite generally, if a creature fails to fit an ecological niche exactly, it follows that that isn't exactly the creature's ecological niche" (*Ibid.*, p. 140).

"Only agents have minds, and only agents act out of their intentions, and natural selection isn't an agent... Surely you may say, nobody could really hold that genes are literally concerned to replicate themselves? Or that natural selection literally has goals in mind when it selects as it does? Or that it's literally run by an intentional system? Maybe. But, before you deny that anybody could claim any of that, please do have an unprejudiced read through the recent adaptationist literature (especially in evolutionary psychology)" (*Ibid.*, p. 122).

"Where we've got to so far: 'sorting-for' is an intensional process. If there is an agent doing the sorting-for, that would account for its intensionality; but, in the case of evolutionary adaptation, there of course isn't an agent. Alternatively, if there are laws of adaptation (laws about the relative fitness of phenotypes), that too would account for the intensionality of sorting-for. But it looks like there aren't any" (*Ibid.*, p. 130).

Appendix I

Other mainstream scientists who avow the "Cambrian Explosion" and/or admit that there is no evolutionary change of some animals in the fossil record

Ernst Mayr—Emeritus Professor of Zoology, Harvard University—one of the original architects of the Modern Synthesis of Evolution—"Paleontologists had long been aware of a seeming contradiction between Darwin's postulate of gradualism, confirmed by the work of population genetics, and the actual findings of paleontology. Following phyletic lines through time seemed to reveal only minimal gradual changes but no clear evidence for any change of a species into a different genus or for the gradual origin of an evolutionary novelty. Anything truly novel always seemed to appear quite abruptly in the fossil record" (Mayr, 1988).

Dr. Jerry Fodor, Ph.D. from Princeton University, and is Professor of Philosophy at Rutgers University—"Insect wings are an evolutionarily significant novelty whose origin is not recorded in the fossil record. Insects with fully developed wings capable of flight appear in the fossil record in the upper Carboniferous (ca. 320 million years ago), by which time they had already diversified into more than ten orders, at least three which are still extant [i.e., still living today, JBG]" (Fodor and Piattelli-Palmarini, 2010, 86).

Dr. Robert L. Carroll, McGill University Professor, Curator of Paleontology, Redpath Museum—"One of the outstanding problems in large-scale evolution has been the origin of major taxa, such as the tetrapods, birds, and whales, that had appeared to rise suddenly, without any obvious answers, over a comparatively short period of time" (Carroll, 1997).

Dr. Stefan Bengtson, Expert on the Cambrian Explosion, Swedish Museum of Natural History—"If any event in life's history resembles man's creation myths, it is this sudden diversification of marine life when multicellular organisms took over as the dominant actors in ecology and evolution. Baffling (and embarrassing) to Darwin, this event still dazzles us and stands as a major biological revolution on a par with the invention of self-replication and the origin of the eukaryotic cell. The animal phyla emerged out of the Precambrian mists with most of the attributes of their modern descendants" (Bengston, 1990).

Dr. Edwin H. Colbert, paleontologist, professor, Columbia University, American Museum Natural History—"Unfortunately, the fossil history of the snakes is very fragmentary, so that it is necessary to infer much of their evolution from the comparative anatomy of modern forms" (Colbert and Morales, 1991).

Drs. Douglas H. Erwin, Curator of Paleobiology, Smithsonian National Museum of Natural History, James W. Valentine, U.C. Berkeley, and David Jablonski, University of Chicago—"All of the basic architectures of animals were apparently established by the close of the Cambrian explosion; subsequent evolutionary changes, even those that allowed animals to move out of the sea onto land, involved only modifications of those basic body plans. About thirty-seven distinct body architectures are recognized among present-day animals and from the basis of the taxonomic classification level of phyla" (Erwin *et al.*, 1997).

Dr. Jeffrey Schwartz, physical anthropologist, University of Pittsburgh—"Although paleontologists have, and continue to claim to have, discovered sequences of fossils that do indeed present a picture of gradual change over time, the truth of the matter is that we are still in the dark about the origin of most major groups of organisms. They appear in the fossil record as Athena did from the head of Zeus- full-blown and raring to go, in contradiction to Darwin's depiction of evolution as resulting from the gradual accumulation of countless infinitesimally minute variations, which, in turn, demands that the fossil record preserve an un- broken chain of transitional forms" (Schwartz, 1999).

Dr. Richard Dawkins, the world's fore- most spokesman for atheism, Emeritus Fellow of New College, Oxford University— "The Cambrian strata of rocks, vintage about 600 million years [evolutionists are now dating the beginning of the Cambrian at about 530 million years], are the oldest in which we find most of the major invertebrate groups. And we find many of them already in an advanced state of evolution, the very first time they appear. It is as though they were just planted there, without any evolutionary history. Needless to say, this appearance of sudden planting has delighted creationists....the only alternative explanation of the sudden appearance of so many com- plex animal types in the Cambrian era is divine creation" (Dawkins, 1986).

Dr. Niles, Eldredge, paleontologist, Cura- tor, American Museum of Natural History, Professor, City University of New York— "Whatever my pattern was showing, it didn't appear to be this mode of expected gradual change... It is the job of all Ph.D. candidates in the sciences to show they can formulate and carry out original scientific research. And, say what one will, there is the unspoken assump- tion that positive results are to be expect-

ed. . . It was no thunderclap, just a plodding, slowly dawning realization that the absence of change itself was a very interesting pattern . . . But Gould has also said that 'stasis is data,' and indeed it is. The trick to seeing stasis itself as a pattern, as a result, and not a nonresult, required only a shift from equating 'evolution' with 'change'" (Eldredge, 1999).

Drs. R.S.K. Barnes, P. Calow and P.J.W. Olive— "Many species remain virtually un- changed for millions of years, then suddenly disappear to be replaced by a quite different, but related, form. Moreover, most major groups of animals appear abruptly in the fossil record, fully formed, and with no fossils yet discovered that form a transition from their parent group" (Barnes *et al.*, 2001).

Drs. Stefanie De Bodt, Steven Maere, and Yves Van de Peer— "In spite of much research and analyses of different sources of data (e.g., fossil record and phylogenetic analyses using molecular and morphological characters), the origin of the angiosperms remains unclear. An- giosperms appear rather suddenly in the fossil record... with no obvious ancestors for a period of 80-90 million years before their appearance" (De Bodt *et al.*, 2005).

Columbia University geoscientist profes- sor, Dr. Arthur Strahler— "Columbia University geoscientist Arthur Strahler wrote that, "This is one count in the creationists' charge that can only evoke in unison from paleontologists a plea of *nolo contendere* [no contest]" (Strahler, 1987).

Bibliography

Ager, Derek V. 1976. "The Nature of the Fossil Record," *Proceedings of the Geological Association*, 87:[2]131-159.

_____. 1981. *The Nature of the Stratigraphical Record.* New York: John Wiley Publishers, third edition.

Allegro, John M. 1986. "Divine Discontent," *American Atheist,* 28:25-30.

Apologetics Press Staff. 1994. "Prophetic Precision," *Reason & Revelation,* 14[12]:96.

Archer, G.L. 1994. *A Survey of Old Testament Introduction.* Chicago: Moody Press.

Austin, S. A., and D.R. Humphreys. 1990. "The Sea's Missing Salt: A Dilemma for Evolutionists," *Proceedings of the Second International Conference on Creationism,* Vol. 2, eds. R.E. Walsh and C.L. Brooks. Pittsburgh, PA: Creation Science Fellowship Inc.

Baker, Sylvia. 1976. *Bone of Contention.* Darlington, England: Evangelical Press.

Bales, James D., and R.D. Clark. 1966. *Why Scientists Accept Evolution.* Grand Rapids, MI: Baker.

Barnes, T.G. 1981. "Depletion of the Earth's Magnetic Field," *Institute for Creation Research Impact,* No. 100.

Batten, Don, ed. 2003. *The Revised and Expanded Answers Book.* Green Forest, AR: Master Books.

Behe, Michael. 1996. *Darwin's Black Box.* New York: The Free Press.

_____. 2005. Personal Communication, July 8.

Bergman, J., and G. Howe. 1990. *Vestigial Organs Are Fully Functional.* Terre Haute, IN: Creation Research Society Books.

Blum, Harold. 1962. *Times Arrow and Evolution.* New York: Harper Torchbooks.

Boyden, Alan. 1947. *American Midland Naturalist,* 37:648-669.

Bradshaw, Robert. 1999. "Tyre," [Online], URL: http://www.biblicalstudies.org.uk/article_tyre. html.

Burgess, S. 1999. "Critical Characteristics and the Irreducible Knee Joint," *Creation Ex Nihilo Technical Journal,* 13:112-117.

Chin K., D.A. Eberth, M.H. Schweitzer, T.A. Rando, W.J. Sloboda, and J.R. Horner. 2003. "Remarkable Preservation of Undigested Muscle Tissue Within a Late Cretaceous Tyrannosaurid Coprolite from Alberta, Canada," *Palaios,* 18[3]:286-94, June.

Clark, Austin. 1928. *Quarterly Review of Biology,* December.

Clayton, John Neil. 1976. *The Source: Eternal Design or Infinite Accident.* South Bend, IN: Published by the author.

_____. 1978. "Does God Exist?" *Book of the Month,* 5[4]:2-3, April.

Clark, LeGros. 1955. *Discover,* January.

Cook, Melvin A. 1970. "Discovery of Human Footprints with Trilobites in a Cambrian Formation of Western Utah," in: *Why Not Creation?* ed. Walter E. Lammerts. Philadelphia: Presbyterian and Reformed Publishing Co.

Coyne, J.A. 1998. "Not Black and White," *Nature,* 396:35,36.

Criswell, W.A. 1972. *Did Man Just Happen?* Grand Rapids, MI: Zondervan.

Darwin, Charles. 1860. Letter to Asa Gray, Sept. 10, in: *The Life and Letters of Charles Darwin,* Vol. 2 (1896), ed. Francis Darwin. London: D. Appleton and Co., p. 131.

_____. 1872 edition. *The Origin of Species,* first published in 1859. London: John Murray, sixth edition.

_____. 1876. In: *Autobiography, The Autobiography of Charles Darwin, 1809-1882. With Original Omissions Restored,* (1958), ed. Nora Barlow. London: W.W. Norton and Co.

_____. 1882 edition. *The Descent of Man and Selection in Relation to Sex,* 2 Vols., first published in 1871. London: John Murray, second edition.

_____. 1971 edition. *The Origin of Species,* first published in 1859. New York: J.M. Dent & Sons.

Dawkins, Richard. 1989. *The Selfish Gene.* New York: Oxford University Press.

Desmond, A., and J. Moore. 1991. *Darwin: The Life of a Tormented Evolutionist.* New York: Warner Books.

Dewar, D., Davies, L.M., and Haldane, J.B.S. 1949. *Is Evolution a Myth? A Debate between D. Dewar and L.M. Davies vs. J.B.S Haldane.* London: Watts and Co. Ltd./Paternoster Press.

DeYoung, D. 1990. "The Earth-Moon System," *Proceedings of the Second International Conference on Creationism,* Vol. 2, eds. R.E. Walsh and C.L. Brooks. Pittsburgh, PA: Creation Science Fellowship.

Dickens, Larry. 2004. "Scientific Foreknowledge," *The Renewing of Your Mind: 2004 Truth Magazine Lectures,* ed. Mike Willis. Bowling Green, KY: Guardian of Truth Foundation.

Dilbeck, W.H. 1978. "Theistic Evolution or Atheism," *Firm Foundation,* 95:3,12.

Dods, Marcus. 1948. "Genesis," in: *The Expositor's Bible,* Vol. 1, ed. W.R. Nicoll. Grand Rapids, MI: Eerdmans.

Dwight, Thomas. 1911. *Thoughts of a Catholic Anatomist.* New York: Longmans, Green and Co.

Elam, E.A. 1925. *The Bible Versus Theories of Evolution.* Nashville: Gospel Advocate.

Eldredge, Niles, and I. Tattersall. 1982. *The Myths of Human Evolution.* New York: Columbia University Press.

Encyclopedia Britannica. 1997. Chicago: Encyclopedia Britannica, Inc.

Enoch, H. 1966. *Evolution or Creation.* Madras, India: Union of Evangelical Students.

Fleishmann, Albert. 1928. *Journal of the Transactions of the Victoria Institute,* 65:194,195.

Fodor, Jerry. 2007. "Why Pigs Don't Have Wings" *London Review of Books* 29 (20): 19-22.

Fodor, Jerry and Massimo Piattelli: - Palmarini. 2010. *What Darwin Got Wrong.* Picador Publishing: New York, NY.

Forsdyke, Donald. 2005. Personal Communication, June 22.

Frair, Wayne, and Percival Davis. 1983. *A Case For Creation.* Chicago: Moody Press.

Free, J.P., and H.F. Vos. 1992. *Archaeology and the Bible.* Grand Rapids, MI: Zondervan.

Futuyma, Douglas, J. 1998. *Evolutionary Biology.* Sunderland, MA: Sinauer Associates, Inc., third edition.

Gee, Henry. 1999. *In Search of Deep Time: Beyond the Fossil Record to a New History of Life* (New York: The Free Press).

Gitt, Werner, and Karl-Heinz Vanheiden. 1994. *If Animals Could Talk.* Bielefeld, Germany: Christliche Literatur-Verbreitung.

Gish, Duane. 1995. *Evolution: The Fossils Still Say No!* El Cajon, CA: Institute for Creation Research.

Goldenhagen, Daniel Jonah. 1997. *Hitler's Willing Executioners.* New York: Vintage Books.

Gould, Stephen Jay. 1977. "Evolution's Eratic Pace," *Natural History,* 86:[5]12-16, May.

_____. 1980. "Is a New and General Theory of Evolution Emerging?" *Paleobiology,* 6:119-130.

_____. 1982. *Evolution Now: A Century After Darwin,* ed. J.M. Smith. London: Macmillan Publishing Co.

_____. 2000. "Abscheulich!—Atrocious!—The Precursor to the Theory of Natural Selection," *Natural History,* 109[3]:42-48, March.

Gould, Stephen Jay, and Niles Eldridge. 1977. "The Return of Hopeful Monsters," *Natural History,* 86[6]:22-30, June/July.

Gould, Stephen Jay, and Niles Eldridge. 1993. "Punctuated Equilibrium Comes of Age," *Nature,* 366:223-227.

Grassé, Pierre-Paul. 1977. *The Evolution of Living Organisms.* New York: Academic Press.

Gribbin, John. 1983. "Earth's Lucky Break," *Science Digest,* 91:36-37,40,102, May.

Grigg, Russell. 1993. "Should Genesis be Taken Literally?" *Creation,* 16[1]:38-41, December.

Hall, M., and S. Hall. 1974. *The Truth: God or Evolution?* Nutley, SC: Craig Press.

Ham, Ken. 1987. *The Lie.* Green Forest, AR: Master Books.

Hammerton, J.A. 1924. *Wonders of the Past: The Romance of Antiquity and Its Splendours,* Vol. 3. New York: G.P. Putnam's Sons.

Harrub, Brad, and Bert Thompson. 2003. *The Truth about Human Origins.* Montgomery, AL: Apologetics Press.

Hayward, Alan. 1985. *Creation or Evolution: The Facts and the Fallacies.* London: Triangle Books.

Helfinstine, R.F., and J.D. Roth. 1994. *Texas Tracks and Artifacts.* Anoka, MN:Published by Authors. Library of Congress number 94-96128.

Hoffman, P. 1979. *Hitler's Personal Security.* London: Pergamon Press.

Holden, C. 1981. "The Politics of Paleoanthropology," *Science,* 213:737-740, August 14.

Hooten, Ernest. 1937. *Apes, Men and Morons.* New York: George Allen & Unwin.

Hoover, Arlie J. 1992. *Dear Agnos: Letters to an Agnostic in Defense of Christianity.* Joplin, MO: College Press Publishing Company.

Humanist Manifesto I and II. 1973. Buffalo, NY: Prometheus Books.

Huxley, Aldous. 1946. *Ends and Means: An Inquiry into the Nature of Ideals and into the Methods Employed for Their Realization.* London: Chatto & Windus.

Huxley, Julian. 1960. *Issues in Evolution*, Vol. 3, ed. Sol Tax. University of Chicago Press.

Jackson, Wayne. 1982. *Biblical Studies in the Light of Archaeology.* Montgomery, AL: Apologetics Press.

_____. 1983. "Our Earth—Young or Old?" *The Restorer,* 3[9]:3-5, September.

_____. 1991. *Christian Courier,* 27:1.

Jastrow, Robert. 1992. *God and The Astronomers.* New York: W.W. Norton and Co.

Ji Q., S.A. Ji, Y.N. Cheng, H.L. You, J.C. Lu, Y.Q. Liu, and C.X. Yuan. 2004. "Palaeontology: Pterosaur Egg With a Leathery Shell," *Nature,* 432[7017]:572, December 2.

Johanson, Donald C. 1996. "Face-to-Face with Lucy's Family," *National Geographic,* 189[3]:96-117, March.

Keith, Arthur. 1979. *Evolution and Ethics.* New York: G.P. Putnam's Sons.

Kerkut, G.A. 1960. *The Implications of Evolution.* London: Pergamon.

Kitts, D. 1974. "Paleontology and the Evolutionary Theory," *Evolution,* September.

The Koran, Surah IX.

Lamont, Corliss. 1991. "Humanism and Civil Liberties," *The Humanist,* 51:5-8, January/February.

Landau, Misia. 1991. *Narratives of Human Evolution.* New Haven, CT: Yale University Press.

Lawton, Graham. 2009. "Why Darwin Was Wrong about the Tree of Life," *New Scientist, 2692:*34-39, January.

Lemonick, Michael D., and Andrea Dorfman.1999. "Up from the Apes. Remarkable New Evidence Is Filling in the Story of How We Became Human," *Time,* 154:[8]50-58, August, 23.

Lewin, Roger. 1980. "Evolutionary Theory Under Fire," *Science Magazine,* 210:883-887.

Lubenow, Marvin. 1992. *Bones of Contention.* Grand Rapids, MI: Baker.

Lyons, Eric. 2003. "It's a Bird! It's a Dinosaur! It's... Archaeopteryx," [Online], URL: http://www.apologeticspress.com/articles/1821.

MacArthur, John. 2001. *The Battle for the Beginning: The Bible on the Creation and the Fall of Adam.* Nashville, TN: W. Publishing Group, a division of Thomas Nelson Inc..

Major, Trevor. 1996. "The Fall of Tyre," *Reason and Revelation,* 16[12]:93-95.

Margulis, Lynn. 2010. In: *The Altenberg 16: An Exposé of the Evolution Industry.* (North Atlantic Books: Berkley, CA).

Marsak, Leonard M., ed. 1961. *Existentialism and Humanism. French Philosophers from Descartes to Sartre.* New York: Meridian.

Marshall, Clyde, and Edgar L. Lazier. 1946. *An Introduction to Human Anatomy.* Philadelphia: W.B. Saunders, third edition.

Matthews, Robert. 1999. "Evolution Research Based On Series Of Blunders," *Calgary Herald,* March 21.

McHardy, G. 1976. "The Appendix," in: *Gastroenterology,* Vol. 2, H.L. Bockus, ed. Philadelphia: W.B. Saunders and Co., third edition.

Moffit, Jerry, ed. 1993. "Arguments Used to Establish an Inerrant, Infallible Bible," in: *Biblical Inerrancy.* Portland, TX: Portland church of Christ.

Moore, Keith L. 1992. *Clinically Oriented Anatomy.* Baltimore: Lippincott Williams & Wilkins, third edition.

More, L.T. 1925. *The Dogma of Evolution* (Princeton, NJ: Princeton University Press).

Morell, V. 1993. *"Archaeopteryx:* Early Bird Catches a Can of Worms," *Science,* 259:764-765.

Morris, Henry. 1963. *The Twilight of Evolution.* Grand Rapids, MI: Baker.

_____. 1970. *Biblical Cosmology and Modern Science.* Grand Rapids, MI: Baker.

_____. 2002. "Things That You May Not Know about Evolution," *Back to Genesis,* No. 160, April.

Morris, Henry and John D. Moore. 1996. *The Modern Creationist Trilogy–Science and Creation,* Vol. 2. Green Forest, AR: Master Books.

Moyer, Doy. 1995. *Standing on Solid Ground.* Russellville, AL: Norris Book Co.

Muller, H.J. 1955. "Radiation and Human Nutrition," *Scientific American*, 193[5]:58-68.

National Academy of Sciences. 1998. *Teaching about Evolution and the Nature of Science.* Washington D.C.: National Academy Press.

Nelson, Ethel R., and Richard E. Broadberry. 1994. *Genesis and The Mystery that Confucius Couldn't Solve.* St. Louis: Concordia Publishing House.

Nietzsche, Friedrich Wilhelm. 1881. *Daybreak,* R.J. Hollingdale translation, 1984 publication. Toronto: Penguin Books.

Nüsslein-Volhard, Christiane, and Eric F. Wieschaus. 1980. "Mutations Affecting Segment Number and Polarity in Drosophila," *Nature,* 287:795-801.

Orr, H.A., and J.A. Coyne. 1992. "The Genetics of Adaptation: A Reassessment," *American Naturalist*, 140:726.

Osborn, H.F. 1918. *The Origin and Evolution of Life.* New York: Charles Scribner's & Sons.

Palmer, Douglas. 2002. "One Great Leap for Mankind," *New Scientist,* 173[2334]:50, March 15.

Parker, Gary. 1994. *Creation: Facts of Life.* Green Forest, AR: Master Books.

Patterson, Colin. 1979. Letter of April 10, 1979 to Luther D. Sunderland, Appalachin, New York, in: *Darwin's Enigma,* 1989, Luther D. Sunderland. Green Forest, AR: Master Books, fourth edition.

Pennisi, Elizabeth. 1997. "Haeckel's Embryos: Fraud Rediscovered," *Science*, 277:1435, September 5.

Pitman, M. 1984. *Adam and Evolution.* London: Rider and Co.

Pfeiffer, C.F. 1966. *The Biblical World.* Grand Rapids, MI: Baker.

Pfeiffer, John. 1961. *The Human Brain.* New York: Harper.

Provine, William B. 1987. "*Review of 'Trial and Error: The American Controversy over Creation and Evolution'* by Edward J. Larson, *Acadame,* 73[1]:51,52, January/February.

Raven, P.H., and Johnson, G.B. 1999. *Biology.* Boston, MA: McGraw-Hill, fifth edition.

Reese, K.M. 1976. "Workers Find Whale in Diatomaceous Earth Quarry," *Chemical and Engineering News,* 54:[4]40, October 11.

Richards, Eveleen. 1983. *New Scientist,* 100:887, December 22/29.

Riddle, Oscar. 1954. *The Unleashing of Evolutionary Thought.* New York: Vantage Press Inc.

Ridley, Mark. 1981. "Who Doubts Evolution?" *New Scientist,* 90:830-832.

Riegle, D.D. 1962. *Creation or Evolution.* Grand Rapids, MI: Zondervan.

Safarti, Jonathan D. 1999. *Refuting Evolution.* Green Forest, AR: Master Books.

_____. 2002. *Refuting Evolution 2*. Green Forest, AR: Master Books.

Sagan, Carl, and Ann Druyan. 1990. "The Question of Abortion," *Parade,* April 22.

Sereno, Paul. 1999. "The Evolution of Dinosaurs," *Science,* 284:2143.

Scadding, S.R. 1981. "Do Vestigial Organs Provide Evidence for Evolution?" *Evolutionary Theory,* 5:173-176.

Schützenberger, Marcel-Paul. 1967. "Algorithms and the Neo-Darwinian Theory of Evolution," in: *Mathematical Challenges to the Neo-Darwinian Interpretation of Evolution*, eds. P.S. Morehead and M.M. Kaplan. Philadelphia: Wistar Institute Press.

Schweitzer, M., and T. Staedter. 1997. "The Real Jurassic Park," *Earth,* June.

Schweitzer M.H., J.L. Wittmeyer, and J.R. Horner. 2005. "Soft tissue Vessels and Cellular Preservation in Tyrannosaurus Rex," *Science,* 307[5717]:1952-1955, March 5.

Schweitzer M.H., J.L. Wittmeyer, J.R. Horner, and J.K. Toporski. 2005. "Gender-Specific Reproductive Tissue in Ratites and Tyrannosaurus Rex," *Science,* 308[5727]:1456-60, June 3.

Scofield, Cyrus I., ed. 1909. *Scofield Study Bible*, revised 1917, reprinted 1998. New York: Oxford University Press.

Sepkoski, Jack J. 1992. "A Compendium of Fossil Marine Animal Families," *Milwaukee Public Museum Contributions in Biology and Geology*, 83:7, second edition.

Shreeve, James. 1996. "New Skeleton Gives Path from Trees to Ground an Odd Turn," *Science,* 276:654, May 3.

Simpson, George Gaylord. 1944. *Tempo and Mode in Evolution*. New York: Columbia University Press.

_____. 1951. *The Meaning of Evolution*. New York: Mentor.

_____. 1953. *The Major Features of Evolution*. New York: Columbia University Press.

_____. 1964. *This View of Life*. New York: Harcourt, Brace & World.

Snelling, Andrew A. 1998. "Rodiometric Dating in Conflich," *Creation,* 20:24-27.

Southerton, Simon. 2004. *Losing a Lost Tribe: Native Americans, DNA, and the Mormon Church*. Salt Lake City: Signature Books.

Spencer, Herbert. 1969. "Illogical Geology," reprint in: *The Works of Herbert Spencer.* Osnabrilck, Germany: Proff and Co., 13:192-210.

Stanley, Steven M. 1979. *Macroevolution: Pattern and Process*. San Francisco, CA: W.H. Freeman and Co.

Stokstad, E. 2005. "Paleontology. Tyrannosaurus Rex Soft Tissue Raises Tantalizing Prospects," *Science,* 307[5717]:1852, March 25.

Surburg, R.R. 1959. "In the Beginning God Created," in: *Darwin, Evolution, and Creation*, ed. P.A. Zimmerman. St. Louis, MO: Concordia.

Szent-Gyorgyi, Albert. 1977. "Drive in Living Matter to Perfect Itself," *Synthesis* 1:1.

Tattersall, Ian. 1996. "Paleoanthropology and Preconception," in: *Contemporary Issues in Human Evolution, Memoirs of the California Academy of Science Series,* eds. W.C. Meikle, F.C. Howell, and N.G. Jablonski. San Francisco, CA: California Academy of Science, 21:47-54.

Teller, Woolsey. 1945. "Evolution Implies Atheism," in: *Essays of an Atheist.* San Diego, CA: The Truth Seeker Company, Inc.

Thompson, Bert. No date. *Issues in Evolution,* [Online], URL: http://www.apologeticspress.org/ rr/ reprints/Issues-in-Evolution.pdf.

Thompson, Bert. 1999. "Causes of Unbelief Part III," *Reason and Revelation*, 19:49-55.

_____. 2001a. *Defense of the Bible's Inspiration.* Montgomery, AL: Apologetics Press.

_____. 2001b. "What's Wrong With Theistic Evolution?" Montgomery, AL: Apologetics Press, [Online], URL: http://www.apologeticspress.org/articles/1990.

Thompson, Bert, and Brad Harrub. 2003. "The Geologic Timetable and the Age of the Earth," in: *Advanced Christian Evidence Correspondence Course: Lesson 4.* Montgomery, AL: Apologetics Press.

Thompson, Bert, and Wayne Jackson. 1992a. *A Study Course in Christian Evidences.* Montgomery, AL: Apologetics Press.

Thompson, Bert, and Wayne Jackson. 1992b. *Christian Evidences.* Montgomery, AL: Apologetics Press.

Thompson, Bert, and Wayne Jackson. 1996. *The Case for The Existence of God.* Montgomery, AL: Apologetics Press.

Turner, R.A. Sr. 1989. *Systematic Theology.* Montgomery, AL: Alabama Christian School of Religion.

Vardiman, L. 1990. *The Age of the Earth's Atmosphere: A Study of the Helium Flux through the Atmosphere,* ICR Technical Monograph. El Cajon, CA: Institute for Creation Research.

Von Fange, E.A. 1974. "Time Upside Down," *Creation Research Society Quarterly,* 11:19, June.

Watson, D.M.S. 1929. "Adaptation," *Nature,* 123:233.

Watson, Lyall. 1982. "The Water People," *Science Digest,* 90:[5]44, May.

Wells, Jonathan. 2000. *Icons of Evolution.* Washington, D.C.: Regnery Publishing Inc.

_____. 2005. "Give Me That Old Time Evolution," *Discovery Institute News,* October 12.

Wiedersheim, Alfred. 1931. *The Science of Life.* New York: Doubleday.

Wieland, Carl. 1994. *Stones and Bones.* Green Forest, AR: Master Books, [Online], URL: http:// remnantprophecy.sdaglobal.org/Librarypdf/Creation-Evolution/Stones%20 and%20 Bones. pdf.

Wieland, Carl. 1998a. "The Mousetrap Man: Interview with Dr. M.J. Behe, The Mousetrap Man," *Creation,* 20[3]:17, June.

_____. 1998b. "The Strange Tale of the Leg on a Whale," *Creation,* 20[3]:10-13.

Wilder-Smith, A.E. 1975. *Man's Origin: Man's Destiny.* Minneapolis: Bethany House.

Williams, Arthur F. 1965, in: *Creation Research Annual.* Ann Arbor, MI: Creation Research Society.

Williams, John G. 1996. *The Other Side of Evolution.* LaVergne, TN: Williams Brothers Publishing.

Wilson, Edward.O. 1982. "Toward a Humanistic Biology," *The Humanist,* Vol. 42, September/October.

Wiseman, J.D. 1974. *The New Bible Dictionary,* ed. J.D. Douglas. Grand Rapids, MI: Eerdmans.

Witham, Larry. 1999. "Darwinism Icons Disputed: Biologists Discount Moth Study," *The Washington Times, National Weekly Edition,* January 25-31, p. 28.

Woodmorappe, J. 2000. "The Fossil Record: Becoming More Random All the Time," *Creation Ex Nihilo Technical Journal,*14:110-116.

Wright, Sewel. 1982. "Character Change, Speciation, and the Higher Taxa," *Evolution,* 36:427-443.

Wysong, R.L. 1976. *The Creation-Evolution Controversy.* East Lansing, MI: Inquiry Press.

Zhou Z., P.M. Barrett, and J. Hilton. 2003. "An Exceptionally Preserved Lower Cretaceous Ecosystem," *Nature,* 421[6925]:807-14, February 20.

CPSIA information can be obtained
at www.ICGtesting.com
Printed in the USA
LVHW01s2323220318
570898LV00003B/3/P